CITIZEN
AND
SOLDIER

CITIZEN
AND
SOLDIER

The Memoirs of
LIEUTENANT-GENERAL HOWARD GRAHAM

M&S

Canadian Cataloguing in Publication Data

Graham, Howard, 1898–1986
Citizen and soldier

Includes index.
ISBN 0-7710-3390-7

1. Graham, Howard, 1898-1986. 2. Generals - Canada –
Biography. 3. Canada. Canadian Army - Biography.
4. Lawyers - Canada - Biography. 5. Toronto Stock
Exchange. I. Title.

FC601.G73A3 1987   971.063'092'4   c86-094609-6
F1034.3.G73A3 1987

Printed and bound in Canada.

McClelland and Stewart
The Canadian Publishers
481 University Avenue
Toronto, Ontario
M5G 2E9

CONTENTS

PREFACE

I t was with reluctance that I assumed the task of writing these memoirs. Only after persistent urging by friends and relations did I finally decide that perhaps they were right and that I should do so. I have had a varied and rather unusual life. It occurred to me that the story of how it had all happened might be of interest and that some of it might even be important and have historical value.

How did a boy from a small farm, of poor family, make his way from being a teenage soldier to the position of Chief of the Canadian Army? That was one career.

How did the boy also make his way to a successful legal practice and to become mayor of his native town of Trenton, Ontario? That was another career.

How did he become, on two occasions, the Canadian secretary to Her Majesty the Queen? That may not qualify as a career, but it was a most honourable and valued experience.

How and why did he come to fill an important national position in the financial and investment community as president of the Toronto Stock Exchange? This, indeed, can be classed as a career, though of less than six years' duration.

All these and other activities to which I refer in the book make up the story of my life. With little exception, I have written from my own experiences and from personal knowledge. Where this is not the case, I name the sources from which I quote.

One thing that bothers me in this project is the need to use the first-person singular – I, me, my – so frequently. It cannot be helped, but I do want to emphasize that in all my undertakings I owe a great debt of gratitude to those many people who helped and supported me in my endeavours.

I would like to acknowledge those who have helped me with this book. They include George Renison, Craig MacFarlane, Jeffery Williams, Ken Smith, Ramsay Derry, J.G. (Jack) McClelland, and my editor, John Robert Colombo. Mrs. Irene Paul and Miss Pearl McCabe of Mississauga Secretary assisted in the typing of the manuscript. It was the late John Easson and his wife Dorothy who initially urged me to undertake the writing of these memoirs.

H.D.G.
Oakville, Ont.

1

URBAN LIFE, RURAL LIFE

When it was announced that I was to succeed Lieutenant-General Guy Simonds as Chief of the General Staff of the Canadian Army, the press release included a short résumé of my background, including the fact that I had been born in Buffalo, New York. Gordon Sinclair, the well-known radio newscaster in Toronto, who was prone to add his own comments to news items, mentioned my appointment and drew attention to the fact that I had been born in the United States. Then he added his own comment, "Too bad they couldn't find a Canadian for the job!"

My first seeing the light of day in an American city came about this way. Both my parents were born in Canada of Scottish ancestry. They moved to Buffalo early in 1895. Three years later, on July 15, 1898, I became their third son. My two brothers, Claude and Ernest, had been born in Canada in 1892 and 1894. My sister, Edith, was born in Buffalo in 1904. In the spring of 1905, my parents apparently decided that their native Canada, rather than Buffalo, was the place where they wanted to raise their family of four. This must have been a soul-searching decision because, by 1905, my father, who started as a motorman with the Buffalo Street Railway in 1895, had risen to the position of inspector, with prospects of continued steady employment at an adequate wage.

My own memories of life in Buffalo up to the age of almost seven years

9

are few but very distinct. I attended Public School No. 24, near Humboldt Park. I remember that it was a new building and within walking distance of our home. Each morning my class formed in the large auditorium with the other classes. We sang "My Country 'Tis of Thee, Sweet Land of Liberty," saluted the Stars and Stripes hanging in front of us, and marched to our classrooms. I recall seeing horse-drawn streetcars still being used on side streets to connect with the newer electric trolley cars, as they were called, on Genesee Street.

My father, as an inspector, travelled by streetcars to various parts of the city and often took me with him. I became quite well known to many conductors and motormen on the cars as Wes Graham's little boy. My widowed maternal grandmother had moved to Buffalo and set herself up as a dressmaker at some distance from where we lived. After a trip or two to visit Grandma with my mother by streetcar, I used to make the journey alone – at the age of five! One day I sat in a streetcar beside a lady who asked me where I was going. I told her my Grandma's address and how I was going to get there. She thought she knew a quicker way, which involved transferring to another car. I should have resisted her advice but I did not, and as a result I got completely lost! However, I was soon put right by one of my father's men and eventually arrived safely at Grandma's place. I had learned my first lesson – beware of advice from strangers.

The Pan American Exposition was held in Buffalo in 1901 and U.S. President William McKinley came from Washington to spend a few days to see it. He was standing in a building known as the Temple of Music when he was shot by an anarchist. He survived for several days and it was thought for some time that he might recover, but it was not to be. The tragedy was on everyone's lips and pictures and drawings were reproduced in great numbers in newspapers and magazines. Although I was just over three years old at the time, I remember learning about the tragedy. When my mother took me to the fair, I was shown the building "where the President was shot." So the event was impressed upon my young mind.

I have happier memories than these. I recall holidays when my parents took us across the Niagara River to picnic at Crystal Beach or Queenston Heights.

In late April, 1905, my father left us and took the train for Trenton, Ontario, to find a new home in his native Canada. A few days later came a telegram telling us to follow – he had found what he wanted. Our furniture was loaded on a railway car, and we followed. The train trip from Buffalo to Trenton took an entire day.

We could have returned to the old homestead in the Township of Murray, northwest of Trenton. This land had originally been bought from the Church of England and settled by my grandfather, Andrew Graham, when he came from Scotland in the early nineteenth century. Instead of returning to his old homestead, my father bought a small farm of thirty acres within the boundaries of the town of Trenton. His reason for so doing was that his children would have a good opportunity for schooling. The old farm was several miles from a public school and several more miles from a high school. The new place was in the north part of Trenton, commonly known as the "Dutch Settlement," and was within three-quarters of a mile of both a public school and a high school. In addition to this, the small farm, if one might call it that, was ideally suited for what he had in mind – namely, market gardening. He wanted to grow crops to sell to the canning factory and to the local stores and at the market.

So it was that, in May of 1905, the family moved from the bright lights of Buffalo, with its paved streets and sidewalks, electric streetcars that stopped within two blocks of our house, indoor plumbing, gas lights on the streets and in the house, and friendly neighbours close by. We moved to a country town, to an area where there was no gas, no electricity, no telephone, no street lights, no paved streets – just dirt roads that were very dusty in summer, very muddy in spring and fall, and very snowy in the wintertime. It took me a few weeks to grow accustomed to the blackness of night. I would look out of the farmhouse window and see not a glimmer of light. Before going to bed, a lantern was lit and we took turns going out to the privy, commonly called the "outhouse," a few rods from the back door. It was only a "two-holer"!

This part of the town was presumably called the Dutch Settlement because the earliest pioneers were from Holland. They still lived on some of the small farms – the Keitels, the Shonikers, and the Orvills. But they had been invaded by the Irish, so we then had in the Settlement the Sheridans, the McAulays, and the Connollys. We must have seemed to the neighbours to be rather a strange family moving into this community because we were, indeed, pioneers. We had absolutely nothing with which to work a farm, but we had other things. For instance, my oldest brother had a bicycle and my older brother and I had tricycles. My mother had a very fine White sewing machine with which she could do all sorts of fancy things and there was not another one in the neighbour-hood. We were well dressed and had a good supply of attractive furniture, including a piano. The neighbours must have thought that we were city-slickers!

But in spite of these fancy items, we had absolutely nothing with which to work a property. We had no stock, no implements, and no tools, and we did have a thirty-acre farm that had to be ploughed, fertilized, and seeded without delay if it was to produce a crop that year. It was now early May, 1905. The place had been vacant and had lain fallow for two years.

My father spread the word to our new-found neighbours and to his many relatives who lived in the vicinity that we would have a ploughing bee on a certain date later that month. I will never forget the morning of the great day because I had never seen so many horses in my life. There must have been thirteen or so teams. The men began to arrive not long after sunrise. Some had ploughs, some had harrows, some had both. Some men brought their wives, who provided great hampers of sand-wiches and other food. My father organized the work of cultivating the fields, establishing the boundaries of what each person would be responsible for with his ploughing, harrowing, or discing.

In no time at all they were all out in the fields doing their jobs. We three boys, all under thirteen, helped mother and the other ladies set up tables on trestles under a couple of apple trees in the orchard. By noon the tables were laden with food of many kinds. I remember somebody had found or borrowed a large school bell. About midday it was rung, and the men, leaving the equipment in the fields, unhitched their horses, brought them in, put them behind the barn, and fed them the hay and grain that they had brought with them for their noonday meal. Soon we had thirty or so people sitting at the tables having a happy time and enjoying a memorable dinner.

I cannot remember whether there was any hard cider on this particular occasion. Certainly there was no whisky, but hard cider was a quite well-known, refreshing drink. No doubt some farmers had remembered to bring a few jugs along with them. After dinner, they returned to the fields, and by the end of the afternoon the whole farm, or at least all that part of it that needed to be, was ploughed, harrowed, and ready for fertilizing and seeding.

For seed and plants, my father went to the Mathews' canning factory, the principal one in town, and signed a contract for a certain number of acres of tomatoes, corn, pumpkins, peas, and beans. The factory provided the seed and plants, charging them to our account.

An early priority, after the purchase of the farm, was the acquisition of a horse. My father was a wise man. He bought a mare, named Polly, which had just foaled a filly we named Olly. Wherever mare Polly went, little filly Olly went also. If we drove to town, Olly gamboled along beside her mother. Often when we stopped in town, Olly (because she was not yet weaned) would have a little lunch from mama. In due course Olly grew to be a fine young mare and we had fun "breaking her" into harness to drive and to ride bareback. We never had a saddle, and many a time I got thrown, but I never suffered more than a sprained ankle.

In time we got other horses and usually kept four. One in particular was a beautiful young mare named Maud. She was a perfect match for Olly, not in colour but in shape and size. I remember Maud and Olly so well because I drove them a great deal for several years. Then I went off to war. When I was in France, I had a letter from Mother saying that the team had been sold to the Army's Remount Depot. From that time on, I watched every team that passed me on the roads of France, hoping we might have a reunion! But, alas, it never happened!

Cows also had a high priority. Father bought two – Betsy had just "freshened," i.e., borne a calf – but in this case we did not claim the calf. Betsy was a good milker and gave us all we needed. But, wisely, Father bought another cow, Nell, that was due to freshen in a month or two. So we soon had two milking cows and, fortunately, a heifer calf. Thus we had milk for the family and cream for our butter.

Pigs were essential if we were to have a supply of meat during the winter, which was six months away, and so a pair of young ones were bought and put in a pen my father built behind the barn. In later years we went into the pig business in a larger way with a sow that produced a litter of fifteen.

I could, and did, have affection for horses, cows, chickens, and their offspring, but I could never work up much love for the pigs! They were such dirty, smelly animals that lived on swill, leavings from the table, and pigweed I had to pull and carry to them in great armsful. But perhaps my feelings were influenced by the way we treated them. Days or weeks after the first litter was born, the males were castrated, and this was done by my father with the help of my brothers and myself. It was a sad and painful experience for me, to say nothing about the ten or so little pigs. So that a quick and neat job might be done, very sharp knives were necessary. I had to turn the grindstone so that my father could make them ready. I always seemed to be the one to turn the grindstone to sharpen knives, axes, scythes, hoes, chisels, etc. The knives being sharpened and all of us dressed in the oldest and dirtiest clothes available, Father and three sons repair to the pig pen. Number 3 son chased around the little pig pointed out by Father, cornered it, grabbed it to his chest, and brought it to Father, who has rigged up an operating table. No. 1 and No. 2 sons helped to hold the poor little pig while Father deftly slit the sac, pulled out the testicles, applied a good covering of unsalted butter to the wound, and fondly, it seemed to me, as though he had compassion for the little eunuch, put it back, squealing like a banshee, into the pen. By the next day the little fellow and those of his brethren who had suffered the same fate seemed to be out of pain and quite happy.

Then, too, there was the slaughtering and curing of one or two pigs in the autumn for our own use. Here again the knife had to be sharp in the point as well as the blade. My father was the executioner, with us to help hold the victim while he felt under the throat for the proper place and with one thrust to the jugular laid the pig low with scarcely a quiver. A great black kettle (we called it a "cooler") was slung on a timber tripod over a heap of firewood and was filled with water. A fire was lit and kept going until the water was boiling hot. With pails we transferred the boiling water to a large barrel, a hogshead, tipped at a slight angle. Two men, my father and the oldest son, took a front and back leg of each dead pig and dowsed it back and forward in the barrel as long as the water was hot. The tendons on the hind legs were opened so that a pointed hickory stick could be passed through them, and by this stick the pig was hung clear of the ground on a sloping rail, resting against the barn. Then, with sharp scrapers, the bristles were scraped off until the pig was quite clean-shaven. Now the knives came into play again and the belly was slit from top to bottom. The "innards" were rescued – kidney, heart – and the intestines were sometimes rendered – boiled – to provide a quantity of fat, and the bladder was always handled carefully so that it could be used as a football.

When all this was done and the insides were thoroughly washed down with warm water, our pig was ready to be taken inside and "cut up." At least one ham, the ribs, and the tenderloin would be kept for early use. The head, the feet, and the hocks likewise would be kept for head cheese and pickling. The rest of the carcass would be cut into parts – hams, shoulders, sides – salted, and carefully put to rest in a barrel for the coming winter. It was all an important part of the life and sustenance of a pioneer family.

The Graham family had to have chickens and one flock was easily started. When Billy Maguire, one of our many cousins, who lived six or seven miles away, came to our ploughing bee, he brought presents for the four children – three hens and a rooster. My little sister got a white Minorca, I got a black Minorca, my older brother Ernest got a Plymouth Rock – all laying hens – and my oldest brother got a rooster, no special breed but very good at his job! We added to this small lot by the purchase of a few young chicks. It was not long before we had an ample supply of eggs.

A great aunt had married a gent who had a carriage factory at Wooler, about eight miles away. He provided us with a buggy – a democrat, a one-horse, four-wheeled light wagon, and when winter came a cutter and light sleigh. At first we had little else except the hand tools father bought at auction sales or on time from local merchants. It was common practice in those days to buy on credit and settle the debts in the fall when crops and produce were sold and when the cheque from the canning factory was received.

When tomatoes were ready for sale, crates that held four pecks were needed. A few good-sized pines on the place were cut down, taken to the nearby sawmill, and cut into pieces for the ends. The sides and bottoms of crates we cut ourselves. When beans were ready for picking, we bought empty sugar bags, usually from the grocers, for a few cents each. The crates and bags were used over and over again for several years, as they would be emptied at the factory and returned to us.

With an eye to the needs of the coming winter, my father had sown timothy, clover, oats, and buckwheat by hand. I can still see him, with a bag of seed slung over his shoulder, striding up and down a field, almost like an automaton, every step the same, every sweep of his arm the same, as he scattered the seed evenly over several acres. When the crop sprouted a week or two later, it was seen to be as smooth and even as though done by a seeder and team of horses. When the 1905 autumn came, and these crops were ready for harvesting, Father again went forth with a scythe for the hay and a cradle for the grain. Again with a sweeping, even, almost graceful motion, he cut the hay and laid it in rows, and with the cradle laid the grain in small loose bundles.

The cradle he used was never used for rocking a baby! It was similar to a scythe except that attached to the blade and supported by the handle was a sort of basket arrangement that supported and collected the grain as it was cut by the blade. For the cornstalks, a sickle was used. Father would grasp the four or five stalks growing in a hill and, with one stroke of the razor-sharp sickle, cut and lay them on the ground. My brother or I followed close behind, with three-foot lengths of binder-twine tucked under our overall bibs, and gathered and tied the stalks into small bundles. These were gathered by my older brother and piled into upright stooks. Scythe, sickle, and cradle were the harvesting implements.

We always kept a plot of beans and peas for our own use and let them ripen on the vines. In the autumn they were pulled up and allowed to dry. Then they were threshed and cleaned for winter use in the kitchen. This was an interesting but somewhat laborious process in those far-off days. When the vines had dried in the sun, they were brought into the barn. About a fifteen-foot-square area of the barn floor was swept clean and then covered with the bean or pea vines, usually well laden with ripe pods. The vines were then thoroughly flailed. We made the flails ourselves using long hickory shafts. The pods would break open and the beans or peas would fall through to the floor. Occasionally we would shake the vines and turn them with a fork to ensure that the pods had been broken open. Finally, after a half hour or so, the vines were forked away and the ripe grain swept up with considerable chaff and put through the fanning mill. This was a large wooden contraption with a crank and a fan, which separated the peas from the chaff and dust. After the operation we would have a few bushels of peas and beans for the winter meals. I can still taste Mother's delicious pea soups, baked beans, and pea puddings. Usually that was a one-course supper with plenty of bread!

In the autumn we stored cabbages and root crops like turnips, mangels, carrots, beets, and potatoes in a root cellar we prepared ourselves. This was simply a rectangular hole in a well-drained location near the house, about five feet deep with a slightly humped roof, well covered with earth and sod. At one end were steps down to a double entrance door. Normally this cellar was frost-proof, but during a very cold spell we would leave two or three lighted lamps burning. A vent at one end of the roof provided the necessary fresh air. In this dreary hole, by the light of a lantern, my brother and I would spend half an hour or more each day after school, slicing roots (sugar beets and turnips) for the livestock or cleaning produce for market or for merchants during the winter or for our own table.

During the first winter, my father built a greenhouse near the barn so that we might grow our own plants from seeds instead of buying them from the canning factory, as we had to do in 1905. This, of course, made more work for us because the pot-bellied coal stove in the greenhouse had to be kept going day and night from early February through April; a very large hogshead had to be kept filled with water near the stove so that warm water would be available to water the seeds after planting and for the small plants. After they were two or three inches high, these had to be transplanted by hand – mostly by little boys' hands – into flats where they would grow under the glass roof and be sturdy and large for outdoor planting in late May. And what kind of plants were these? The variety and quantity of produce were surprising. We grew tomatoes, onions, celery, peppers, cabbage, cauliflower, lettuce, and a variety of garden flowers in smaller volume. Other vegetables, such as beans, peas, turnips, beets, and potatoes, were planted directly outside and not started in the greenhouse.

During our first year on the farm, we lived in much the same way as our grandfather and his grandfather before him had lived more than a hundred years before, using primitive methods of sowing and reaping and threshing. The few necessities that we did not produce ourselves came mostly from trading our produce with merchants in town for salt, sugar, white flour, and a few other necessary staples. What few clothes we needed were mostly made by Mother or Grandma, who came to stay with us from time to time.

2

LIFE ON THE FARM

During those first few months, I learned a great deal about life on a farm, including the facts of life. One day I went with a neighbouring boy, two or three years older than my eight years, to fetch home his cows from a pasture about a mile away. They would be taken there in the morning after being milked and brought home in the late afternoon for milking and for the night. One of the cows in the herd of about fifteen was acting as I had never before seen a cow perform. It continually tried to mount the back of another cow and push it off the road. My friend did not explain this phenomenon, so at the supper table, with Mother, Father, Grandma, my sister, and two brothers all enjoying their meal, I gave them an exciting account of the cow's playful game. My brothers sniggered with their hands over their mouths and almost choked. My father gave them a frowning look and then said, "Howe, that cow wants to have a family, and since it can't talk that's the way it tells people it wants to get married." I suppose that was a fair explanation to an eight-year-old, but the next morning my brothers, being twelve and fourteen and well versed in the ways of nature, laughed at my innocence and set me straight. "Ha, ha! That cow doesn't care about a family! What it wants is to have a good bull," they exclaimed. In any event, one of our own cows soon exhibited the same desires, and this time I saw the consummation of its lust. In a very short time I knew all about the functions of bulls, boars, rams, roosters, and stallions.

I learned, too, on the farm that some living things had to suffer and many had to be killed. I learned, also, that if an animal or bird died a natural death, it was not fit to be eaten. Consider, then, my reaction to the first time I saw my father lay the head of a hen (which I had caught for him) on a block of wood and with one swipe of a hatchet sever the head and then lay the body of the bird on the ground. Its legs twitched and its wings flapped in reflex action, and I cried out, "Kill it, Pa, before it dies!" I thought that was quick thinking on my part. But he explained that it was already dead and that these spasms were caused by nerves. My brothers heard my cry and for years teased me by saying, "Kill it, Pa, before it dies!"

Frogs and barn pigeons sometimes met their fate at our hands because they were a great delicacy. We had a lady cousin who came to visit us once or twice a year – dear Maggie. As soon as she arrived she would say, "Well, boys, how about a pigeon pie?" Mother had too many more important things to do. We would easily catch by net half a dozen, common barn-variety birds, chop off their heads, and pluck and clean them. Cousin Maggie did the rest in a large sauce pan and hot oven.

If it was springtime, she would say, "How about a feast of frogs' legs for breakfast?" The three of us, armed with stout sticks, would go down to the creek that ran through our farm and, within half an hour, would knock off a couple of dozen good-sized frogs. We would make sure they were dead by hitting their heads smartly on a rock, cut off their hind legs, skin them, and deliver them to Maggie. Next morning, for breakfast, we had a real gourmet meal.

Even though we had a great deal of work to do connected with the gardening business and barn chores, the Graham boys seemed to find time for hobbies. One of these was raising rabbits. We frequently sold them and made a few quarters or half-dollars. Another project was raising pigeons. We started first with a pair of fantails. They were reddish in colour, and when they spread their tails like little peacocks and strutted around the barnyard, they were a sight to behold. Then we got a pair of pouters. I have not seen any of that breed since those boyhood days. The pouter was a larger bird than the fantail. It was bluish in colour, and its great pride and joy was to blow its breast out to the size of a softball, so large indeed that it could scarcely see to its front. Pouters seemed to want to put on a better show than their fantailed cousins.

Next we got a pair of tumblers. These birds seldom walked about the yard. When not in flight, they sat on the roof or a perch well above the ground. Their great forte was to fly in tight circles to the blue sky until they were scarcely visible, and then tumble almost to the ground in backward somersaults. Finally we got the homers, or carrier pigeons, which we trained to fly "home" over great distances. We would give one to the baggageman on the train and he would let it out, first at Brighton, eight miles away. Next week he would let it go at Cobourg, thirty miles away, until finally we had them flying from our old home, Buffalo. We had a good loft over the stables and, of course, plenty of food. The birds would mate, lay, sit, and hatch three or four squabs, but they never crossbred. From selling the birds we also made a few shekels.

Within a week or two of arriving at Dutch Settlement, my older brother Ernest and I were introduced to the public school. My oldest brother, Claude, though only thirteen at the time, was kept busy at home and never did attend a Canadian school. Nevertheless, he became a very successful businessman, running at different times Ford and General Motors dealerships. He owned a large block of downtown property and for many years was a member of the Trenton municipal council.

Gradually we acquired such basic farm equipment as a horse-drawn seeder to replace the hand-sowing of grain and a cultivator to take the place of much hoeing. (The weeding of long rows of vegetables still had to be done by hand.) A threshing machine was hired to replace the flail and fanning mill; a mower reduced the use of a scythe; a binder relieved the cradle. But as labour-saving equipment came along, my father rented additional property, like the fifteen-acre place next to our lot, two fields nearby for hay and grain, and a large field of pasture across the road from our property. More cattle and horses were acquired. For a year or two we sent milk to the cheese factory but then, apparently realizing that more money could be made selling the milk than having cheese made, my father bought a milk route complete with wagon sleigh, large cans, and the necessary "measures."

The "measures" came in pint and quart sizes. In those days bottles and cartons were not used; the customers left their pitchers or pails on their doorsteps and inside each was a cardboard ticket, red for a quart, blue for a pint, or cash for the quantity of milk required. We had our own tickets printed with the words "Graham's Dairy" on them. It was always a little source of amusement that the painter, in repainting the wagon, printed on one side "Graham's Dairy" but on the other side got his spelling confused and printed "Graham's Diary." My father never bothered to have it changed. Claude really looked after the milk route, except for washing the cans, which was my dear mother's lot. We had several cows but still had to buy many gallons of milk from nearby farmers. Each evening two or three of us boys would go, after supper, with the democrat in summer and the bobsleigh in winter, to collect this milk; frequently some of the neighbour boys would tag along.

Each winter in the early 1900s, one or more stage stock companies would come to town and put on plays like *Peck's Bad Boy* and *Uncle Tom's Cabin*. Occasionally we would have a spare quarter dollar, the price of a gallery seat, and go to see one of these shows. There were no moving-picture shows in these days. *Uncle Tom's Cabin* was a great favourite, for good reason. In the play, the devilish villain is a white man named Simon Legree, a sadist of the worst type. He carried a black snake whip with a long lash, and his favourite amusement seemed to be flailing poor old, silver-haired Uncle Tom with this whip. Uncle Tom, in shirt sleeves, would hold his hands above his head, and Legree would wallop him around the waist and chest with the whip, which would make a crack almost like a pistol. Uncle Tom would flinch and squirm as if in great agony.

Well, the Graham boys wondered about these manoeuvres and how Uncle Tom ever stood the punishment. Since we had a whip with a long lash we could crack like a pistol shot, my older brothers decided to experiment on their youngest brother, me. The day after the show, when we were all doing chores in the barn, I took off my reefer (a short warm jacket) and sweater coat and stood in shirt sleeves with arms raised. Claude – a kinder brother never lived – gently lashed me around the waist. No pain. Gradually he increased the treatment to a very hefty wallop. Still no pain, but a resounding crack. We had solved the problem of how Simon Legree lashed Uncle Tom night after night without damage.

A day or two later, when we went to fetch the milk from nearby farmers, the two neighbouring Hicks boys were with us. It was a beautiful winter evening, cold, with a full moon shining on new-fallen snow. We told them how Claude could thrash me with the whip and I could take it without a whimper. Of course, they did not believe it. So when we were almost in sight of our barn, having collected seven or eight large cans of milk in the bobsleigh, we said, "We'll show you that Howard can take all Claude can give." Why we did not wait until we got to the barn, I will never know. We stopped good old Ned, usually a docile horse. There was no other traffic on the quiet country road, and the five of us stood at the rear of the sleigh. I took off my reefer and sweater and put them with the milk cans. Claude, with such an interested audience, put on a good show. He gently flicked the lash around my chest, beneath my upstretched arms. Then bending to his task, he flailed me with all his might, and with a mighty smack of the whip. At the sharp crack of the lash, old Ned shot off like a bat out of hell. We could hear the milk cans banging and the sleigh bells ringing as he made straight for his barn, with the five of us scared stiff and racing after Ned, the milk, and my reefer and sweater. It took only a few minutes for us to reach the barn, and there stood Ned, trembling and snorting from fright. No milk was spilled, but some cans had been knocked over and were somewhat dented. Those milk cans had caps that fitted six or eight inches into a sleeve at the top of the can, somewhat like a cork in a wine bottle. Lucky for us, none of the caps had come out of its socket.

Winter was the season of "deep freeze." It set in usually about mid-November, and many feet of snow fell from November to March. Trenton is located at the mouth of the Trent River where it empties into the Bay of Quinte, which, in turn, is a long, narrow, fifty-mile reach of Lake Ontario. The Trent and the Bay, being fairly shallow, from fifteen to thirty feet deep, and protected from most of the high winds, would freeze over early in the season to a safe depth for skating and ice boating, horseracing, and, later, ice-harvesting (the waters being almost devoid of pollution). When the snow piled up, the skating and ice boating were curtailed, but the ice-harvest could be carried on by scraping the snow from a few acres of ice. In the early days of the century, ice in large quantities had to be stored for use during the long and usually hot summer days. The cakes of ice were cut by hand with a saw with a long blade and a handle at one end. To harvest ice was a somewhat dangerous and a very arduous job indeed. But it provided a few dollars for those of us who had horses and sleighs and would make contracts to fill ice houses, of which there were a great many. Electric refrigeration was not yet in vogue, and ice dealers had to store large quantities to meet their customers' needs. It was a rare winter, indeed, when some of the men cutting and hauling the large and heavy cakes of ice from the water to the sleighs drawn up at the edge of the hole did not slide into the icy depths.

Ice boating was very popular and exciting. It is said that an ice boat is the only vehicle that can travel at the speed of wind without added driving force. To have one's skates on and to hold a rope tied to the rear of an ice boat travelling across the glassy surface at a speed of twenty-five or thirty miles an hour was exhilarating but also hazardous. Many skinned noses, knees, and shins were the results.

Horseracing on the ice, with two-wheel buggies (gigs) or high, specially built racing cutters, was a favourite sport for many farmers in the area. As soon as it was safe, a quarter-mile track was marked out on the Trent River by implanting evergreen bushes in the ice and snow. Horse-drawn ploughs kept the snow clear of the track, leaving a keen icy surface. By chance, or more likely by design, the start and finish points for the races were just behind the Prince George Hotel, and there was always a well-cleared path to the bar room, which had a back entrance to the river. Very convenient! I have no proof but it was possible that the Prince George underwrote the cost of laying out the track and paying the stakes. The stakes were small, perhaps $15, $10, and $5 for first, second, and third in some races. The "big ones" were $25, $15, and $10. Still, the races were a gala occasion and good sport for a small market town on Saturday afternoons during the winter.

No motor cars were available to carry us to distant skiing or to resorts for winter pastimes. I never saw a ski until many years later, but we did have snowshoes. Hockey was always a popular winter sport, and Trenton usually had and still has good-to-excellent junior and intermediate teams playing in leagues with nearby towns. In the large pasture, which my father had rented for several years, there was a swampy area of three or four acres edged on one side with a row of cedars. Each autumn the swamp flooded, and when the frost came we had a splendid area for skating just across the road from our house. This became widely known as Graham's Pond, and we boys made a snow plough that enabled us, with the held of old Ned, to keep the pond in good skating condition for the whole winter.

With the help of pals, we built a good shelter in the cedars along one side of the pond, and with the use of stumps and boards made seats for the many dozens of visitors to use when changing boots to skates. We also cut and split cedar rails from nearby fences and usually had quite a cosy fire burning in the centre of the shelter. Why we never burned the place down, I will never know. In the spring the fence, from which the rails had been taken, had to be repaired. But my father never seemed to complain.

We had sleds and snowshoes for winter activities. When the road was well packed with snow and sometimes icy, we had many a happy time speeding down a long, fairly steep hill only a short distance north of our house. Snowshoes were both for utility, in getting about after a heavy snowfall, and for pleasure, when we wanted to join a group from church or school, particularly on a Sunday afternoon, for a hike across the winter fields. None of these diversions interfered with our never-ending chores of cleaning stables, feeding and watering livestock, milking, rubbing down horses, sawing, splitting and piling wood, shovelling paths to the road, to the barn, to the greenhouse, and, most important, to the privy!

Winter evenings were never a bore. True, we had no television, no radio, no telephone, no stereo. But usually there was homework and we always had something to read by the light of one or two oil-burning lamps, as we all sat around the kitchen table with a wood-burning kitchen range a few feet away. Behind it was a well-filled box of hardwood. Filling it was one of my jobs. The reading material included *The Boys' Own Paper* and the Henty, Alger, and Scott books received as Christmas or birthday gifts. *The Youth's Companion*, a weekly magazine, a daily newspaper, either *The Toronto News* or *The Mail and Empire*, all came into the house. My father read the newspapers from beginning to end and was considered well versed in Conservative Party politics. For several years my oldest brother gave me, as a Christmas present, the subscription to *The Boys' Own Paper* published at No. 4 Bouvrie St., London, England. I mention this because I enjoyed the *BOP* so much that, on my first leave to London in World War I, I made it a point to go to that address and see where this great magazine, filled with stories of British adventure and exploration, was printed.

Three other "annuals" were studied with great care by both parents. *Rennie's* and *Steele Briggs* were seed catalogues received in December, and Eaton's mail-order catalogue came in early autumn. My mother, in particular, thoroughly studied the Eaton's book, and in due course sent off a lengthy order, after the cheque from the canning factory was received, for pants, underwear, overcoats, sweaters, stockings, Christmas gifts, and such other items that she or Grandma did not knit or sew, as mitts, scarves, socks, toques, aprons, and dresses.

Father, in his turn, studied the seed catalogues, and in early January mailed a long and quite costly order for the types of seeds that he knew from experience would yield the sort of produce for which our soil was most suitable and which was also most attractive to the customers. I still remember some of the names – Yellow Golden Danver Onions, Giant Pascale Celery, Danish Baldhead Cabbage. In addition, he would add one or two fruits or vegetables that were not so well known. One year, I recall, he tried peanuts. But we had no luck. The summer was too short, but we kids learned that peanuts do not grow on a tree or a bush. Another time he tried salsify, otherwise known as oyster plant. This is a root vegetable, much the same as parsnip but with a distinctly oyster-like taste when boiled and served with a cream sauce. The customers did not take to salsify so he discontinued it. I mention these facts to indicate that my father seemed to have an inquiring and inquisitive mind, and though he had little formal education, he was skilled in many trades.

As I mentioned, ripe peas and beans came from the fields and, after the flailing and cleaning by means of the fanning mill, were ready for winter use by the family. We did not lack a good variety of root vegetables and fruit. Our house, built in the mid-1850s, was a two-storey frame structure with a typically large farm kitchen of one storey. At one corner of the kitchen was a large pantry, and before the winter frost set in the pantry shelves were loaded with home-grown, home-cooked, and canned raspberries, gooseberries, currants, rhubarb, and crocks of pickled cucumbers, cauliflowers, onions, and the like.

The original owners, from whom he had bought the farmhouse in 1905, had wisely planted an orchard of about forty trees surrounding the house. I say "wisely" because the variety of fruit was quite exceptional. Apples included the Yellow Transparent, which could be eaten before September; the Duchess, an early cooking apple; the Snow, excellent autumn eating; the Tolman Sweet, quite a good keeper and most tasty when steamed; the Northern Spy, for all-winter raw eating and cooking; then there were the trees of pears, Flemish Beauty and Bartlett, excellent for canning and raw eating, and several trees of plums and cherries, all good for preserving. Strangely enough, there was also a row of about eight hickory trees along the lane from the road to the barn, and these produced the largest hickory nuts I have ever seen. We gathered them when ripe and put them on the woodshed roof to dry so that the husks could be easily peeled off. There was what we called a "hickory nut orchard" along the Trent River, about a mile from our place, and in the fall we would go there and gather a bushel or two of nuts, though these were smaller than our own.

In addition to the hickory nuts, we gathered bushels of hazelnuts from unused pasture land a couple of miles from home. Popcorn was another winter treat, and of course we grew our own. In the fall, when the cobs were ripe and dry, we braided them into bunches of ten or twelve and hung them in the barn. During the winter, we would strip two or three ears from the braided bunch, and the kitchen was livened by the machine-gun crackling as we worked the wire popper on the hot, kitchen-range wood fire.

And so the winter evenings were enjoyed cracking nuts – hickories had to be done with a hammer and one of mother's flat irons because they were too hard for a nut cracker – popping corn, munching apples, doing homework, reading, playing checkers, croquignole, or dominoes, all in the light of one or two kerosene lamps, until perhaps eight-thirty or nine o'clock. It was early to bed because six o'clock in the dark of the morning was the time to be up and out to do chores before breakfast.

I have tried to describe our way of life at the turn of the century, and have mentioned my father but have said almost nothing of my mother. It is not easy to find words to describe her complete and utter dedication to her family and her home. From May, 1905, to May, 1909, we were a family of four children, aged from one year to thirteen years, a maternal grandmother in good health, about seventy years old, and Father and Mother. We lived in a farmhouse built not long after the Battle of Waterloo. We had no central heat, no pumped-in water, no electricity, no inside toilets. It was all very primitive, with no modern labour-saving devices such as washing machines, electric irons, hot-water heaters, and the like. Consequently, there was a lot of hard work to be done in the kitchen and pantry.

My mother's work started early and never ended. First on the day's agenda was breakfast, invariably oatmeal porridge or corn meal, home-baked bread, home-made butter, home-made jams, and sometimes home-produced eggs, with a few slices of home-smoked bacon. By the time this feast was ready, Father was back from early chores at the barn and with pails of fresh milk. It was Mother's task to put the fresh milk into large flat pans – the process was called "setting the milk" – and to put the pans in the coolest place available in the pantry. Next morning the cream would be skimmed off the pans and put in earthen jars and left to stand until it was ripe for churning. At first we had a crock churn, about three feet high and one foot in diameter, with a large opening at the top with a wooden cap into which fitted a wooden piston or plunger. Being the youngest son, I was designated Mother's helper, and many an hour I sat on a chair, legs astraddle the churn, pumping that piston up and down until the butter appeared, sweet and creamy. After a year or two, as with outdoor equipment, Mother got a barrel-type churn, which again I operated quite easily by pushing a handle back and forth, which rocked the "barrel" on its axis and produced the butter in larger quantities and probably somewhat quicker than did the crock.

The butter being produced, Mother had to "work it" with a wooden ladle in a large wooden butter bowl, to press out the milk, add salt, and, when at a suitable texture, press it into a wooden mould that held exactly one pound. Sometimes, as I recall, she would turn out five or six pounds from a churning and hence have a few pounds for "trading" at the grocery for perhaps sugar or spices or flour. The buttermilk left after the churning was much favoured. In the summer, when we went to the fields for the morning stint, we took, when available, a two-quart pail of buttermilk with a chunk of ice in it. A few swallows of that brew was a rare treat on a hot summer day!

I cannot remember seeing Mother sit down for breakfast, but she must have had a snack "on the go," after she set the milk and cleaned up the breakfast dishes and sent the family off to school or out to the fields. Then it was upstairs with slop pail (awful name!) to empty and clean chamber pots, which as adult males we seldom used because we answered the call of nature in the outhouse just before going to bed and first thing after rising in the morning. (This is not the nice subject to include in one's memoirs, but I hope to remind a later generation how their forebears lived.) After beds had been made (and I never remember going to rest in an unmade bed) and the upstair rooms cleaned, the lamps, usually two or three, were brought down to the kitchen table. Along with two or three downstairs lamps and usually a lantern, the chimneys were cleaned, the wicks trimmed, and the oil replenished. Then all were ready for the night's use.

On Monday morning, the weekly washing was done. In the winter, with pairs of long underwear, heavy shirts, long stockings, etc., this was no light task. The wood fire in the kitchen stove had to be roaring. The wash boiler required fairly warm but not very hot water from the "reservoir" (a tank attached to the back of the stove that held about two and a half gallons). The boiler fitted quite neatly over the two front lids of the stove and was soon boiling. Into it went the dirty clothes with soft soap or shaved flakes from a Comfort bar. Near the stove were two tubs, one with hot water, one with cool. After the required time, and after prodding and swirling, the clothes, usually using the broom handle, were lifted to the hot water tub. Here rested the washboard and more soap used as needed. The laundress scrubbed the clothes as she bent over the washboard. Then, for rinsing, they were put through the ringer and into the cooler tub. I was Mother's helper in these operations, usually before I went off to school in the morning. I brought water as needed from the soft-water cistern outside, drawing it up by a pail on the end of an eight-foot stick with a hook on the end. I turned the handle on the ringer as Mother fed the clothes through the rollers.

In the winter it was often a problem to get the clothes dry, and some would have to be hung in the kitchen on temporary lines. All these operations, plus the cooking of meals, made the kitchen a pretty moist and smelly place. In the winter, a thick coating of frost made the windows almost opaque.

A day or two after the wash came the ironing. Again I had to see that a good hot fire was kept blazing and there was a plentiful supply of wood in the box behind the stove, so that after I was off to school Mother would not have to carry it in from the woodshed. Like everyone else in these days, we had a set of four or five flat irons, with one removable handle which, as one iron got cool and was put back on the stove for reheating, would be snapped off the cool one and snapped on a hot iron. It was a very simple operation, taking only a second or two, and since the ironing board (home-made) would be set up resting on the backs of two chairs near the stove, little time or heat was lost when Mother "changed irons."

Baking, in addition to the normal cooking of three meals a day, was another major operation. I think it was done on Wednesdays and Saturdays. Bread dough was mixed and kneaded the night before and left to rise overnight in the large mixing pan. In the morning the heap of dough was carved into proper sizes, shaped, dusted with flour, and put in the loaf-size tins. Again, a roaring woodfire had heated the oven, and in due course a supply of the staff of life was ready for the family. Pies, cakes, cookies, and puddings my mother produced in large quantity for regular meals and for "piecing" between meals. In the winter, the heat of the stove was usually welcome during these washing, ironing, and baking efforts. But I can tell you that in the heat of summer the farm kitchen was as hot as Hades.

It was Mother's responsibility, sometimes assisted by Grandma, to keep husband and children well clothed and in good health. In the winter-time, I recall being dosed with warm onion syrup, which was usually kept in a small bowl on the back of the stove. Also during the winter, a dose of wormwood was administered each morning before breakfast.

What this was for I do not know, but it probably helped to keep colds away.

One problem during the winter was to keep one's feet warm and dry. We did not have the rubbers and overshoes that are available for children today. Perhaps we could not afford them, for we had only leather-laced ankle boots and sometimes moccasins or shoe packs. We used beef tallow to grease the boots, but wading through snow as we had to do almost daily made it impossible to keep the feet dry and warm. At night the footwear would be put on the stove front or set at the edge of the oven in the hope that in the morning they would be dry. Then, if they were dry, they would be hard and stiff. We wore stockings that were above knee-length until we got out of knickers and into long pants at about age sixteen. Stockings would be hung at night on a chair-back close to the stove. If I seem to dwell on the foot problem it is because any painful memories I have of winter life between the ages of five to ten are of cold, wet, chilblained feet that on more than one occasion were frozen. Some relief from chilblains came from cutting a potato in half, rubbing the exposed portion white with salt, and then rubbing the salt into the sore soles of the feet. This treatment helped but did not cure.

In the long winter evenings, everyone relaxed except dear Mother. When the supper dishes, pots, and pans were cleaned up, by washing them in a very large pan with hot water from the stove reservoir and kettle, she sat with us around the kitchen table with the clothes that needed mending or with knitting needles clicking. When these mundane tasks were finished, she would produce her little basket with the crochet hook and thread and work away on lace mats or lace for edging lingerie and other articles. It is not surprising that I never saw her harness, hitch, or drive a horse, milk a cow, feed chickens, or do any outdoor chores. She kept her "operations" within the realm of household work. I suppose that most women of those days did all the things my mother did and raised even larger families. Yet I have read many autobiographies in which little is said about the essential work of The Mother.

We had music of a sort in the family circle. When Claude was about sixteen, he joined the town band and brought home a rather large brass horn with three keys called an "alto." Ernest, then fourteen, played a Jew's harp and a mouth organ, as many other kids did, and now bought a piccolo. Mother could "chord" on the piano most of the old hymns and some of the old songs like "Where, Oh Where Is my Wondering Boy

Tonight" and "The Maple Leaf For Ever." On a Sunday afternoon, usually in the summer, we would gather in the parlour, which was not used in winter, and go through a repertoire of old-time hymns like "Nearer My God to Thee," "Rock of Ages," "Abide with Me," and a great non-hymn favourite, "The Old Oaken Bucket that Hung by the Well."

In time Claude graduated to the clarinet and Ernest to the flute. The trio of mother and two sons produced quite tuneful but rather mournful music. Grandma thought I should join the party, so for my tenth birthday she bought me a violin. I drew the bow across the strings a few times, but the noise was so dreadful that neither I nor the rest of the family could stand it. I never did play the instrument. At age sixteen, my mother did succeed in inducing me to take piano lessons, as my sister Edith did, from Sister Mary Gerard, a kind, gentle Astercian sister in the small Roman Catholic convent in Trenton. I was not an apt pupil, sorry to say, and was always a bit chagrined at the better progress made by my sister, who was six years my junior. The piano lessons lasted only about a year, and a few months later I enlisted in the army.

If Mother wanted to go to town to shop or attend church or make a visit, Father or one of us boys drove her. One of the annual visits she made was to see Aunt Kate, one of her father's sisters, who lived in Colborne, a village about twenty miles west of Trenton. She would take the train for Colborne on a Friday. On Sunday morning, two of us, usually Claude and I, would leave home before seven in the morning with the buggy and drive the twenty miles to Colborne. The buggy was what we called a "runabout." It had no top or canopy; it had steel tires and an open back so that it could carry several bags and boxes. The roads were unpaved and in the summer heavy with dust. It took us about four hours to reach Aunt Kate and Uncle Ed's place. He was a beekeeper and had an extractor to separate honey from wax from his own and neighbours' colonies of bees. After having dinner at noon and an early supper with Aunt and Uncle, we three would enjoy the cool of the summer evening on the long drive back home. The buggy seat, being large enough for two, meant that I sat on my mother's and brother's knees in turn or in the back with legs hanging out.

When we became more prosperous, we added a second buggy, also produced in Uncle Bill Anderson's factory at Wooler. It was built for both appearance and comfort in that it had rubber tires, light-coloured cloth upholstery, and a shiny black chair-shaped seat for two. In the old or the new buggy, we would take Mother on a Sunday to visit Aunt Jess at Frankford, just seven miles north of Trenton, or Aunt Mary at Wooler, eight miles away. Such visits were the only holiday Mother ever had during those early years.

Mother seldom did work outside the house, but she did sometimes help in preparing the vegetable loads for market. This was a major project at the height of the season. For the large Saturday market – the markets on Tuesdays and Thursdays were smaller – we would start early on Friday morning to pull root vegetables, dig potatoes, pick tomatoes, beans, peas, cucumbers, squash, and corn, cut celery, cabbage, and cauliflower. Wheelbarrows were used to bring the produce to the edge of a patch and then to put it into the democrat or onto a stone boat to take it to a shaded spot under some large apple trees and near the well. While Father and a son were pulling and hauling, other sons assisted with help from the neighbours (at twenty-five cents a day) were washing, peeling, or stripping the produce to make it attractive. Tubs of water from the nearby pump were used. Strangely enough, we peeled off the yellowed outside skins of large "black-seed" onions and washed them so that they were pure white. Vegetables like carrots, beets, and onions were tied in bunches after being washed and having the tops trimmed back. It was in the bunching and tying job that Mother sometimes helped. This work, which took from early morning to late evening on Fridays, was really a thankful change and almost a rest, compared with the weeding, on hands and knees, the hoeing, cultivating, and other arduous tasks performed in the heat (sometimes more than 100 degrees Fahrenheit) and dust of the open fields. This was how we spent our summer holidays!

Near the shady trees where we were preparing the produce for market stood the market wagon. This was a handsome vehicle made, of course, by our Uncle Bill Anderson at Wooler. It had a painted and varnished chassis, with shelving that projected over the wheels, and a light tan canvas top with side and back curtains that could be rolled up. It was only a "one-horse shay," but it could take a very large load. Crates of tomatoes and potatoes would be put on the floor of the wagon. Then boards would be stretched across from the shelving over the wheels, on which the tied bunches of vegetables were piled almost to the roof. We aimed at a sort of variegated colour scheme. Bright yellow carrots were piled high, next to them the clean white onions, then the dull red beets, and beside them parsnips and radishes or a pile of sweet corn. Cauliflow-ers were stuck on spikes protruding from a six-foot board that, when we got to market, was propped against the back of the wagon. Cabbages were treated the same way and propped on the other side of the wagon.

All these preparations were made the night before market day so that on Monday, Wednesday, and Friday evenings, the wagon sat ready for driving to the Trenton market early the next morning. And it had to be very early. For some reason, the town council would not rent a permanent place to any market gardener. It was a case of first there for the best stand. We usually were in time to get the same place week after week, which was important so that regular customers knew where to find us. At the height of the season we were often sold out of some type of produce before noon, when we went home. We would often take orders for items in the canning category, such as tomatoes, cucumbers, pickling onions, and cauliflowers, to be delivered the following week.

Mother, in addition to doing all the housework, was the one who sold the produce on the market. I can never remember my father taking any part in the operation. Other farmers' wives did the same thing, and I think Mother thoroughly enjoyed the change from household labour. She made many dear and lasting friends who were her competitors on the Trenton market. Ninety per cent of the buyers were women, and she would tell some of the buyers about a new recipe for sweet pickles, rhubarb tarts, chili sauce, or johnny cake.

We were always trading butter or eggs or produce for the few groceries we needed to buy. I have a sad recollection of my experience in one of those trades when I was nine. One morning, when new green peas were at their best, Mother said, "Howe, I wish you would pick a peck or two of nice peas and we'll shell them and you can take them down to Mr. Auger for some white sugar. I'm just about out of it." She produced the basket, and in short order I had it filled. She and I sat under the spy apple tree and shucked about two quarts of peas. Then she said, "Oh! I've got a dozen nice fresh eggs that you can take. I don't think it will be too much sugar for you to carry home." She put the peas on paper in the bottom of a covered market basket, but paper over them, and put a dozen eggs on top of the peas.

Off I went down the road, happy to have the trip to Mr. Auger's grocery store about a mile away, whistling "The Old Oaken Bucket that Hung by the Well," the hot dust of the road feeling good between my bare toes. I had gone only a little way when an empty lumber wagon overtook me, the driver slowed down, and I nimbly put the basket over the tailboard and hopped onto the reach that projected about two feet behind the tail. The wagon had no springs, and the steel tires rumbled and bumped over the rough country road, even rougher in town where the streets were not paved and crosswalks caused heavy bumps. As we came to Mr. Auger's little grocery store, I hopped off, grabbed the basket, and went in to do the trading. Mr. Auger was a kindly man who had come to Trenton from Quebec thirty or forty years before, with a large number of other French-Canadian families whose descendants still live in Trenton. They had come to Upper Canada to work with the Gilmore Lumber Company. Mr. Auger opened the basket, but instead of a dozen nice white eggs he saw a very good imitation of an omelette. Not all the eggs had broken but the innards of those that had ran down onto the peas. It resembled a green pea omelette! Mr. Auger, being a kindly man, took the basket to the kitchen behind the store, salvaged the unbroken eggs, and with dippers of water rinsed off the peas and restored their natural colour and spread them on papers to dry. The upshot was that I got the white sugar with only a pound or two "docked" because of breakage.

Home I went, this time walking all the way. When Mother saw my parcel, she was disappointed that Mr. Auger had not allowed more sugar. I was honest and explained the reason. All she said was, "Well, Howe, let that be a lesson to you. Eggs don't ride easy on a lumber wagon." I'm still rather timid with eggs!

The old farmhouse in which we lived was as comfortable as most. There was no cellar because it was built on rather low, poorly drained land. To keep drafts from around the floors in the winter, we drew loads of manure from the barnyard and piled it about three feet high all around and tight against the house. This did not introduce an unwelcomed smell because it was frozen and frozen manure is odourless. But the senses of smell and sound were actively titillated in these times in farming areas, in a way seldom experienced by most residents in either urban or rural communities today. Some of the odours were sweet and pleasant – fields of clover and of buckwheat in blossom, and new-mown grass, and apple blossoms. But if there were pleasing odours, there were also some dreadful "stinks." The stables, of course, both in town and in the country, yielded some pretty thick smells, especially the pig pens. In the house one was accustomed to the odour on clothes used in the stables and among the livestock. But in the kitchen and throughout the house, the smell of fat being rendered from freshly killed pig's intestines, as they simmered in pots on the stove, almost "took the cake." Head cheese in the

33

making was also, to my little nose, a putrid scent, though somewhat deadened by the free use of various spices. But, believe it or not, sauerkraut in the "working stage" was the most awful of all. We always made a large hogshead for ourselves and for selling or trading. It was prepared by shaving a layer of cabbage into the barrel, then adding a generous sprinkling of salt and so on until the large barrel was filled. On the top was then put a loose wooden head with a large stone on it. The hogshead had to be kept in a warm place for several weeks, as the salt and cabbage worked together to ferment and produce that delicious German dish. The warm place in our house was behind the kitchen stove, and that smell was "out of this world." Still, sauerkraut was another source of revenue, as we traded or sold it to the grocers.

As there were smells that are no longer sensed, so there were sounds that are seldom heard today. Most people, in town or in the country, had a rooster or two. At the crack of dawn, these cocky birds began to advertise their resources with a crowing both loud and clear. I used to think that they were not only trying to impress the females in their harem but also trying to out-crow their neighbouring sheiks. Their rousing voices were soon followed by the wild birds, of which there were great numbers of many varieties. It was the birds that kept the insect and vermin population under control. The fruit trees were never sprayed in our little orchard, nor did other crops need the chemical treatment they require today. We sprayed the potato field with Paris Green, but people with smaller patches simply flipped the little black and yellow bugs into a pan or pail with a stick.

As the sun rose a bit, there were the usual sounds of fowl and stock. The chickens were proudly cackling after delivering their daily eggs. The horses were whinnying for their breakfast, the calves bleating for their milk. In the winter even before dawn, the sleigh bells began to sound as neighbours were on the road to town for work or business. And what a lovely collection of bells there were. We had several single bells to strap on the underside of a horse's collar, and one set of very tuneful chimes – seven bells riveted to a wide strap; the largest bell was in the centre and the three on each side in different sizes, each rendering a different tone. Then there was the string of about forty small bells on a strap that was buckled round the horse's middle. On some cutters and sleighs, three or four bells were fixed to the underside of each shaft. Strangely enough, one soon got to know the sound of everyone's bells, so that we could tell from inside the house who was passing on the road. Sometimes, when we did not "get" the sound, we wondered who the stranger was! Another pleasant winter sound was the singing of the telephone wires. On a clear, cold winter night, the sound was as clear as the tunes of an enormous harp that never paused. Where we were living the Northern Lights, too, would snap and crackle as their shafts shot like great icicles over our heads.

In the spring, as soon as the ice disappeared from Graham's Pond and other low-lying areas, the frogs came forth from their winter holes and soon set the evening air vibrating in a resounding symphony of song. It seemed almost as though the frogs practised their various parts. There were the high tenors, the deeper baritones, and at what was surely a solo part, the deep bass of the old bulls would cut in. The singing of these ugly-looking amphibians was welcome in the warmth of a spring evening in the otherwise quiet of the countryside. Late summer and early autumn brought the crickets and their evening's "orchestra of strings." To me it was always a pleasant, peaceful finale to a long day of toil.

But of all the sounds, in all seasons, I must mention one that struck a most welcome note. It was man-made, the Angelus bell rung from the steeple of St. Peter-in-Chains Roman Catholic Church on a slight hill in the centre of the town. The walk to St. Peter's was better than a mile, but as the crow flies or the sound of the bell carried, it was much less. The Angelus bell called the faithful to prepare to commemorate the Annunciation at six a.m., noon, and six p.m. The early ringing was seldom heard, but at noon and late afternoon it was the signal to pile tools or implements and make for the kitchen. It went: Boom–boom–boom– pause –boom–boom–boom– pause –boom–boom– pause, and then ten or twelve steady rings. The bell was not very tuneful, having a very flat note, but its clang carried clearly over the rooftops and across the fields to summon us from our toil at noon and at the end of the afternoon.

So much for the sights and smells and sounds and varied activities we experienced at Dutch Settlement from May, 1905, to May, 1909. Then there was a change.

3

FARM WORK, SCHOOL WORK

The twenty-fourth of May was the Old Queen's Birthday. It was a holiday from school, but not from work on the farm. My oldest brother, as usual, had delivered the milk and on his way home had squandered twenty-five or fifty cents on firecrackers as a treat for his younger brothers. So it was that after noon-day dinner, we had a half-hour of fun before going back to work in the fields. We lit and threw the crackers as high as we could into the air, and this was done close by the house. May of 1909 had been an unusually dry month, and the old curled, cedar shingles on the house were as dry as tinder. Apparently one of the firecrackers had exploded on the roof of the house. We noticed nothing, and after the short period of fun, off we went to continue our work in a nearby field. Soon we heard from Mother, back at the house, shrieks of "Fire, fire! The house is on fire!" We dropped everything and ran for the house. Sure enough, a good pillar of smoke was already rising over the roof and some flames were showing.

Claude jumped on his bike and raced off down the rough and dusty road to the fire station more than a mile away in town. We did not have a phone, nor did our neighbours. In a remarkably short time, considering the distance and the fact that the fire engine was horse-drawn and the firemen, all volunteers, had to be summoned from their various jobs by the clanging of the fire bell in the tower of the fire hall, the fire engine and hose cart appeared. But alas! They were of no use because one well had already been pumped dry, and our neighbour's well was too far away for the available hoses. In the meantime, helpers from far and near, who could see the smoke, had rushed to the scene. Old and young, male and female, carried out furniture or threw bedding from upstairs windows. Others brought ladders and buckets and formed lines that passed the water-filled pails to the men or the ladies. What a scene in the usually peaceful surroundings, where apple blossoms were just breaking forth, where we had expected to continue to spend many happy if arduous years.

But we were lucky. Scarcely a breath of air was moving. The fire was confined to the roof and main part of the house. The one-storey kitchen was not damaged. The furniture, clothes, and bedding were salvaged. And no one was hurt. Relatives, friends, and neighbours were generous and helpful. The cause of the fire was never discussed, except to assume that it had started from a spark. We did have a wood-burning kitchen stove, but I was sure that the firecracker was responsible.

Before nightfall we had moved the rescued furniture to the barn, where most of it stayed for the next six months. Mother, Father, and baby sister, age five, were put up for the night by a neighbour, and we three boys slept in a tent, loaned by a neighbour and hastily put up before dark. In this tent under an apple tree, we would sleep from May 24 to well into November of 1909. On June 24, exactly one month after the fire, my mother gave birth to her fifth and last living baby, another son. Up to that time, Mother, Father, and little sister had been sleeping in a corner of our large kitchen, while we boys were in the tent. A neighbour provided Mother with a bedroom, where her boy baby was born. When I went to see her, she said, "Well, Howe, you've got another brother. What should we call him?" I was almost eleven years old and had been reading an enthralling book, *The Scottish Chiefs*, all about Robert and Bruce and others of his ilk, so I said, "I think Bruce would be great." Bruce it was.

While we boys slept in the tent, Mother, Father, baby Bruce, and little Edith lived and slept in the corner of the kitchen. But this arrangement could not continue for more than a few summer months, and our parents had to decide without delay whether to repair or rebuild the old house or to build a new one. Father decided to build a new house on our land. One or two houses in town had been built with concrete blocks, a new type of construction that appealed to him. He determined also that any new house must have a good dry cellar with lots of head room – he was six feet

37

tall – and be of modern design with a large cistern under the kitchen, which would have a sink with a hand-pump capable of forcing water up to a tank in the attic. This would be the source of water for bath and toilet. All of these plans showed my parents' desire to improve and modernize our home. Electricity and the internal-combustion engine were still not available to us.

The site for this modern "mansion" was not to be in the lovely orchard because that area was too low for a deep cellar and for proper drainage for the proposed bath, toilet, and sink equipment. So it was that the site chosen was on a knoll, some five hundred yards south of the old house and with scarcely a tree in sight, except for a few rather scruffy cedars. It was not an attractive location, hot in summer and cold in winter and a quarter of a mile from the barn. Although in time the planting of trees and hedges improved the house area, the long walk to the barn to do the chores was always a nuisance.

The new house was completed in mid-November, less than six months from the time of the fire. We were able to move in not a day too soon, as we boys were still sleeping in the tent and winter was upon us. A good job must have been done because to this day, more than three-quarters of a century later, the house stands sturdy and straight as the day it was built.

The CPR about 1911 built a new railway line from Montreal to Toronto and fixed Trenton, one hundred miles east of Toronto, as a divisional point where engines would change, new crews would take over, freight trains would be reassembled, and rolling stock would be checked. The roundhouse and other service buildings were located only half a mile north of our little farm. My father wisely decided that these CPR activities would result in a need for housing and he built two nice houses in our old orchard area. Had he lived, he probably would have built several more. During these ventures we still ran the real business of gardening, although my older brother, who later became a pharmacist, did stay out of school for one year. We had no motor equipment but did, about 1912, have the house wired and serviced with electricity and had a telephone installed. We even got a hand-operated cream separator so that mother was relieved of the task of setting the milk and skimming the cream.

When the new house was built, Father had arranged for an outside entrance to the cellar as well as one from the kitchen. Through the outside entrance we brought, in the autumn, a large quantity of vegetables for sale during the winter – cabbages and celery, which we could keep until Christmas, and root vegetables, such as potatoes, beets, carrots, and turnips. The cellar had a dirt floor. Each fall we put a barrel of apple cider in the cellar, which in due course became hard and finally turned to vinegar. The barrel lay on its side with the bung up, so that from time to time the barrel could be easily opened and the contents tested. We would invite our teenage friends to have a few snorts of hard cider. Very refreshing but not palatable! To make the operation feasible, we kept a handful of long wheat straws on a beam in the basement, and through these we sucked the elixir from the barrel until it was out of reach of the straws. Toward spring Father went down to move the cask, and on giving it a mighty heave he almost fell on his face because it was so much lighter than he had expected. He may have surmised the answer but never commented except to say, "That damn bung must have been loose and allowed evaporation!"

It was necessary for the CPR to off-load livestock being transported in slat-sided railway cars after they had been en route for twenty-four hours or even less, and feed and water the animals, clean the cars, put in new straw bedding, and reload them for the onward journey, which in this case was usually to Toronto. At the south edge of their assembling yard, the railway built a series of fenced enclosures or corrals, the gates of which opened off a central laneway. The company then contracted with a merchant, Esli Marsh, in Trenton, to provide hay, grain, and straw and to arrange for the necessary help required to do the job of off-loading. Mr. Marsh needed a reliable, experienced, and readily available man to take on this task and induced my father, a friend of long standing, to accept this part-time job. It was the most miserable, dirty, difficult, back-breaking effort you could ever imagine. I was then fourteen years old and, when not in school, took my share in this operation. I have no idea what money my father made, but whatever the amount it was well earned. Nowadays trains travel faster, cattle spend fewer hours on moving from here to there, and from what I observe most of them are shipped by truck. I doubt that the laborious and dirty job (which went on in the cold and snow of winter as well as in the heat and rain of summer) I had a part in is any longer necessary.

It might be assumed that with all the activities of farming, house-building, and railway work I had little time for school, but that was not so. From 1905 to 1915, when I graduated from high school, I scarcely missed a day. True, I was up early and did my allotted chores, and even some work in the fields in the busy early summer and early autumn months, but Mother always saw to it that I washed and changed from overalls to knickers and was off to school by 8:40 a.m. in time to get there before last bell at nine o'clock.

There were two public schools in Trenton, one on each side of the river. I started in May, 1905, at the Trenton College Street School on the east side and only a fifteen-minute walk from home. But on reaching "Fourth Book" (seventh grade) in 1910, I had more than twice as far to go to the Dufferin Street School on the west side of the river because the College Street school taught only up to "Third Book" (sixth grade). For some reason unknown to me, the schools in Ontario were organized not on a grade basis, as they are now, but on a "book" basis. There was a Junior First Book and a Senior First Book, a Junior Second and a Senior Second, and so on to the Junior Fourth and Senior Fourth. Then you tried your entrance examination for high school, with the emphasis on "tried." There was no promotion on merit or the "year's standing." If you were an average student, you had no problem passing through a junior and senior "Book" in one year, with the exception of Fourth Book, which invariably took two years.

Politics had a part to play in education. I was in Senior Fourth in 1911, the year of the great reciprocity issue. This was a proposal supported by the Liberal government under Sir Wilfrid Laurier for closer trade relations with the United States, by which each country would make concessions favouring the importation of certain products from the other. The Conservative opposition in the House of Commons, under Sir Robert Borden, opposed the Laurier policy and favoured closer trade with Great Britain. One day in class, our teacher, P.W. Fairman, who was also school principal, left the class, as he frequently had to attend to some pressing matters. During his rather short absence, the twenty-five of us were left to ourselves. Almost at once two pupils, Art Hayes of a staunch Tory family and Don Fraser of a loyal Grit family, started arguing and shouting at each other about the Tory bastards and the Grit s.o.b.'s, and in a minute or two they were battling each other up and down the aisle and over desks. Then the door opened and Mr. Fairman walked in! The noise ceased, and the two culprits were summarily sent home and suspended from school for two weeks. In the riding of South Hastings we took our politics seriously, even in our junior years.

In June, 1911, just short of my thirteenth birthday, I passed my entrance exams to high school and then started there in September. I always found learning to be easy, and my four years from 1911 to June,

1915, when I passed both my junior matriculation and normal-school entrance examinations, were almost but not quite uneventful. I took all subjects that were available, including Latin Grammar, Latin Authors, French Grammar, French Literature – I mention these two subjects because in future years both were to be most useful to me. I was a member of the Cadet Corps and played the usual games of that day – soccer, baseball (softball had not yet been invented), and tennis (when I had the time, which was not often).

In 1914 my father let me off work to go to Cadet Camp, held for all Cadet Corps in eastern Ontario at Barriefield Camp near Kingston. I well remember those ten days as a wonderful holiday. The year previous, in July, 1913, our corps had won the district cup for general appearance and proficiency. (By July, 1915, many of those boys were in the trenches in France.) We wore khaki uniforms, puttees, red epaulettes, and broad red facing on the cuffs of the tunics with brass buttons. We made a good showing on parade when the Governor General H.R.H. The Duke of Connaught, brother of King George V, mounted on his charger with mounted staff behind, took the salute on a march past. It was generally agreed that in 1914 we were as good as our Corps had been in 1913 when we won the cup. But in the meantime, the Williamsburg High School Corps had found a sugar-daddy sponsor, who had fitted them out with kilts (Black Watch, I think), red jackets with white cuffs, and white spats! They really were a fine-looking corps with half-a-dozen pipers to lead them by the reviewing party. They won the cup!

But I must tell you of an occurrence during my first year at high school that almost ended my academic career. Between the two classrooms there were folding doors, which, when opened, provided an assembly room. A weakness in this arrangement was that, when the folding doors were closed, one could hear what was going on in the other room, especially if there was any unusual sound, such as a teacher raising his or her voice slightly above normal. One afternoon, we in my room could hear the principal, Robert Whyte, in the other room, expounding in a very loud and angry voice, which he was wont to do when he could not get the right answers from stupid or inattentive pupils.

On our side of the folding doors, we were having a lesson in history from a recently arrived, very attractive young lady teacher, who was exceptionally quiet of voice and mild-mannered. I am afraid some of us young pups took advantage of her and on this particular day, I, being seated near the back of the room, was sliding a one-foot wooden ruler up

41

the aisle to a young lady near the front, who was sliding it back. At the same time, another stupid boy had a flexible wooden ruler he was snapping on his desk and thereby making a resounding crack.

Suddenly the door from the hall was flung open and there appeared Mr. Whyte. His face was red with anger, his short white moustache seemed to be bristling. "Who is playing with that ruler?" he fairly shouted. Well, I had fun playing with a ruler so I stood up, alone. "Come up here, Graham," he ordered, and I went with no delay. When I stood before him there was no delay either; he simply hauled off and gave me a slap on the side of my face that sent me reeling, and said, "Go into my office." I left the room in a hurry but not for his office. I simply went down the stairs in about three jumps and ran all the way home.

Mother said, "What brings you home an hour early?"

I replied, "Mr. Whyte hit me and I'm not going to school anymore," and went out to the fields behind the house to be alone and think of the great injustice I had suffered. It was not my ruler sliding on the floor like a whisper that Mr. Whyte had heard, but Max cracking his ruler on the desk that upset the principal.

About five o'clock that afternoon, a buggy drove up before our house and in came Mrs. Pattee. Dear, dear Mrs. Pattee was a widow, an elderly teacher of French and English literature, who was not only a great teacher but was also a gentle and kindly lady, respected and adored by her pupils. Mrs. Pattee and my mother knew each other, and Mrs. Pattee said at once, "Has Howard come home, Mrs. Graham?"

My mother answered, "Yes, he is upset and says Mr. Whyte hit him." Mrs. Pattee then explained what had happened and said it would be a great pity if I left school because I was a first-class scholar and should have a good future in one of the professions. Perhaps if she talked to me she could "smooth things over." My mother heartily agreed and called me in from the back, where I was nursing my wounded pride.

I really liked school (it was a great relief from the grind at home) and it did not take long for Mrs. Pattee to convince me that I should appear at school the next morning and nothing more would be heard about the unfortunate incident. So back I went and that was the case.

In those days the theory and practice of distribution of wealth through the medium of taxation by the government was not in vogue, so we had to make our way through life "on our own," so to speak. This was the practice both at home and at school. We bought our own school books. If there was a small school orchestra, the members provided their own instruments. If we wanted uniforms for sports, we bought our own. An exception was the cadet uniform, which was provided on loan by the Department of Militia and Defence. The first cheque I received in my life was for five dollars. It came from the Department of Militia and Defence as a reward for taking instructions and passing an examination in semaphore signalling with flags. I used the money to buy a second-hand bicycle.

About 1913, the members of the school baseball team decided that they would like uniforms. We found that we could order white shirts and knickers and black-and-yellow stockings (our school colours) and pennants to fasten across the shirt fronts with yellow "T.H.S." on a black background for about five dollars apiece. But who, in these times, had five dollars for a ball uniform? Not many of us. Then some smart young man conceived the idea that we might make a contract with the school board to remove the winter's accumulation of ashes, which was heaped behind the school by the janitor. In previous years the contract had been let to various teamsters in town. We needed a team and wagon, but that was easily provided by the Graham boys. My father was always co-operative in helping us with any of our projects, whether ploughing snow from the pond or hauling the bandwagon rented from the livery in town to take the team to neighbouring towns for games. We landed the contract, we moved the ashes, we raised the money, and in short order we had magnificent new uniforms.

Shortly, we went to Frankford to play against the local team. They had no school diamond and the game was played in the local fair ground, which was normally used as a cow pasture. I played centre field and we were doing famously when a long fly ball was hit to my territory and I made off at high speed to capture it. Alas! Alas! In watching the ball, I missed seeing a really great flat fresh cow deposit, very slippery indeed. I landed and slid on the seat of my clean white knickers. Where the ball went I do not know. I was dreadfully embarrassed, and even after a scraping job by my pals, I carried the sight and smell of cow manure until I got home. We did not carry a change of clothes on those trips. My dear mother did her best, but I regret to say I carried a light stain in the seat of my pants evermore!

On August 4, 1914, Canada and our Mother Country – as we then thought of Great Britain – were at war with Germany. I first heard of this not from radio or television, of course, but from a neighbour as he drove home from town where he had seen the newspapers. I had just turned sixteen and had no thought of enlisting, but two boys, Gerald and Burwall Hicks, aged sixteen and seventeen, enlisted within a few weeks. In those days, if you were big enough, you were old enough to enlist. Late in April, 1915, the same neighbour who told us that war had started shouted to me as he drove past, "The Hicks boys are missing, Howard." And so they were, with never a trace, along with thousands of others who were lost in the terrible carnage at St. Julien and Langmark, where poison gas was first used by the Germans.

After Labour Day of 1914, it was back to school for the last year. Needless to say, the war, and the prospect of it lasting until I would be old enough to enlist, was in my mind as it was in the thoughts of most of my classmates. This thought of enlisting was greatly abetted by the attitude of the populace, which was almost entirely of Anglo-Saxon lineage and hence strongly in support of their British forebears and relatives still living in the old country. Frequent recruiting meetings were held in halls and in Waller's Opera House, often on Sunday evenings. I recall vividly our minister, who came from Scotland, the Rev. Geordie Ross, cutting short the evening service so the congregation could go in a body to the Opera House. Here we would be regaled by flights of oratory condemning the "Unspeakable Hun" and Kaiser William, who had initiated the dreadful conflict. About mid-1915 we began to hear at these recruiting meetings from officers who had been wounded and returned home. Strangely enough, their stories of life in the trenches and of the deadly perils of war seemed to increase the desire of many of us to get into the fight, rather than deter and frighten us from the possible results. It was in this atmosphere of war and excitement and intense patriotism that I continued and finished my years at high school and continued to work at home with no clear idea about the future, but with the certain hope of eventually joining the army. I had not yet reached my seventeenth birthday.

In February, 1915, the high school put on a "concert," as it was called. There was much more than music in the program. One item was a demonstration of drill by a selected number of members of the Cadet Corps. I was a sergeant and one of the performers in this exercise. In addition, I had a much more important assignment, namely to make a speech, the subject being "A Plea for a New High School." I had to prepare this *magnum opus* myself and took the whole matter very seriously. About 1912, a new public school had been built on the west side of the river. I got the name of the architects in Toronto who had designed and been in charge of that building and wrote them about my task, asking if they could give me a rough estimate of the cost of a new high school with classrooms – laboratory, assembly hall, and basement gymnasium – to accommodate about two hundred pupils. They were very kind and sent me a very rough estimate – I think it was $110,000. I then consulted the town clerk-treasurer about the means and methods of financing such an enormous project, as it seemed in the days. He was very helpful and gave me my first lesson in high finance. He explained that "debentures" would have to be issued for the amount of money needed for the school. They were something like a mortgage on a farm, which I knew about, but the debentures and interest on them would be paid by taxes levied against all the owners of property in the town. He thought that perhaps these debentures could be sold "over the counter," to use his expression, to people and businesses in town, so that a commission would not have to be paid to a broker or underwriter.

As a result of these inquiries, I developed, I think, a quite well-reasoned speech in which I first described – perhaps with some slight exaggerations! – the poor conditions then prevailing in the high school located in the upper storey of the building that served as a public school on the ground floor. I then dealt with the estimated cost and methods of raising the money and paying the cost of a new building over a period of twenty years. I then had my favourite teacher, Mrs. Pattee, read it over as a critic. She thought it was first-class, but toned down some of my more lurid statements. Then I memorized the twelve-minute effort and practised speaking it aloud, mainly behind the barn. So it was that on the night of the concert, after taking part in the drill exhibition in my crisply pressed uniform, it was my turn to expound. This I did according to plan, with only one ad-lib phrase. As I was holding forth on the lack of proper heat in the winter, and how snow actually drifted in through cracks in the walls of the school, I must have built the drifts a bit high because I had to stop a minute as the result of loud laughter from my audience. I then said in a stern, sixteen-year-old voice, "This is no laughing matter, the snow does drift in." And it did, of course, but only a little and in one place!

My efforts in preparation and delivery in the wide-open spaces behind the barn paid dividends because the local newspaper in its report on the concert made this effusive comment: "After Miss Hazard's solos, the sensation of the evening took place in an address by Howard Graham entitled 'A Plea for a New High School'. . . . The young orator (and he is an orator worth listening to) made clear the disabilities of the present building. . . . " The published reference went on to give quite a complete account of my address. There was an interesting sequel. I returned from overseas in May, 1919, and a few days later was invited to the high school closing dance being held in the auditorium of the new high school. Perhaps my "plea" nudged the town fathers into action.

The year 1915 brought many changes in the life of our family. My father, who had suffered from ill health, became much worse in mid-summer and was able to do very little work. Claude married in early autumn and set up housekeeping in one of the houses we had built three years before in the old orchard. Ernest, having graduated from high school in June, 1914, and being unfit for military service because of lameness, was apprenticed to a local druggist in preparation for attending the University of Toronto's School of Pharmacy.

We discontinued taking produce to market but sold it wholesale to local grocery stores. There were many of these because the day of the large chain store had not arrived. People had to shop within walking or short horse and buggy distance from their homes. More help than in previous years was hired to harvest the canning factory crop. When the root crops were ready to pull, we kept enough only for our own use, and instead of putting the balance in the cellar, we sold tons of carrots, beets, potatoes, onions, turnips, and parsnips to Graham's Dehydrating Factory. I drove the team with wagonloads of such produce the seven miles along the Trent River to the factory in Frankford. It was owned by another Graham whose son, Robert Graham, is now a retired army colonel and an Ontario provincial judge.

While we did most of the fall harvesting, my father was making extra money by taking contracts to pack apple crops. Trenton was and still is the centre of many large apple orchards. In most cases the crop was packed in barrels for shipment, usually to the British market. Father, having been raised in this apple country, was skilled at picking, sorting, and packing the fruit, and also adept at judging the quality and quantity of apples in an orchard simply by walking through it. Hence he would make a deal with the owners, most of whom he had known as a boy, to pick, sort, and pack the crop, either at a fixed price or at a price per barrel. He then recruited gangs of five or six men each and for six weeks in the fall did a quite lucrative job.

My oldest brother learned the tricks of the trade and, in the fall of 1915, when Father was very ill, he did much the same as Father had done. Having finished high school, and apart from helping to get the crops harvested, I had the odd few days when I would go with Claude to an orchard and help in his operations. My job was usually to bring from the farmer's barn the barrels, which had been delivered there from the cooperage mill, out into selected spots in the orchard. I did this with a horse and democrat. This task took only two or three hours, after which I quickly learned to sort the apples on the folding canvas tables set up at strategic points in the orchard.

Apple trees then were much larger than modern trees, and quite long, pointed ladders had to be thrust up into the branches by the pickers, who used a special type of rounded wicker "apple basket." Although my brother himself did the "tailing" and "heading" and pressing of the barrel, I soon learned the knack of tailing and heading. To tail a barrel, you selected evenly sized apples, clipped the stems short, and leaning into the barrel placed these apples with stem end down in a circle around the bottom so that the apples touched the side and touched each other snugly, but not so they would be bruised. The circle had to be completed so that no apple could be easily moved. Other concentric circles were added until you came to the centre where the proper-sized apple was placed so that all the fruit was locked in and immovable, and so the barrel was properly "tailed." One might think this was an easy operation, but in fact an amateur could fiddle around for half an hour or more to get the right fit and appearance.

Once tailed, the barrel was filled with apples which had passed over the sorting tables, but they were handled gently to avoid bruising. The barrel being filled almost level with the top, it was "headed." This was a more difficult process than tailing because you had not a completely level surface such as the bottom of the barrel to work on. Apples were placed in circles with stem end up. They would be perhaps two inches above the barrel edge. The top for the barrel, usually in three pieces, was placed on the fairly level circles of apples. The press, which had two slightly convex prongs a little longer than the barrel was deep, was hooked to the bottom of the barrel and centred over the loose pieces of top. The handle was turned gradually, and the head of the press pushed the apples down so that the head pieces were level with grooves in the barrel staves. The top hoop of the barrel was gently tapped into place, thus tightening the staves and securing the barrel head.

This may seem very complicated, and it was. My father had acquired the skill as a young man, before he moved to Buffalo in 1895, and my brother learned from him. I was quite good at tailing a barrel, but never got to the stage of packing.

In November, 1915, my father learned that he had cancer and only a few weeks to live. On December 11, he passed away, at the early age of fifty-four, leaving my mother with my older brother Ernest, my sister Edith, age eleven, younger brother Bruce, age five, and me, age seventeen, at home. I will not dwell on those days of sadness. We older ones made a good Christmas for the youngsters and, of course, all rallied round Mother and the home, as a good Scots clan should do.

At once I got a job as a clerk-assistant bookkeeper in the office of the Trenton Electric and Water Co. Ernest got some remuneration from the druggist with whom he was serving his apprenticeship. We had a good CPR engineer as a tenant in the third house we had built, and we rented out the land. Claude kept horses and livestock for some time. And so we made our way, as others did in these days, without the aid of government or service clubs and other organizations that now do so many good works.

During my last year at school, I served as the local reporter for the Belleville *Ontario* newspaper. How I got into this racket I cannot now remember, but it led to my getting many invitations to concerts, weddings, and other social events, and I was paid at the rate of a few cents for each line that appeared in the paper. I naturally made the most of every event and was somewhat miffed when the editor of the *Ontario* curtailed some of my material. I found time also to work in a men's clothing store on Saturdays. I was paid one dollar a day, which was taken out largely in trade. My work was mostly refolding shirts, socks, underwear, and the like and putting them back in place after the two men clerks had displayed the goods to customers. As well, for some months I worked in Webster's Grocery on Saturdays, serving a few customers but mostly weighing up sugar, salt, rolled oats, and other groceries that came in barrel lots and had to be bagged in different weights of two, five, and ten pounds. Many customers phoned in their orders, and when the other two employees were busy, I would take these orders, fill them, and have them ready for the delivery wagon rounds.

It was at this point in my life that the road I travelled made a sharp turn.

4

MY FIRST WAR

March 6, 1916, was a Monday. I was wakened by Mother, as usual, who tapped the stovepipe that came up from the stove below through the floor of our bedroom and thence to the chimney. This pipe brought the only heat we had in the bedroom. I rolled from bed and looked out the window. It was just breaking day, and heavy snow-laden clouds were so low they seemed to be touching the tops of the tall, gaunt maple trees that lined the back end of the farm, about a quarter of a mile away. It had not snowed during the night because the trodden path that ran from our back door across the field to the barn was clearly visible. I dressed in my Sunday best (the first long-pant suit I had ever owned) and went down to the kitchen. I did not have to shave because I was fair of face and had hardly a bit of fuzz.

Mother was having a cup of coffee. She had already had breakfast with my older brother, who had gone off to his work in the drug store, having to walk the mile or more and open the store before 8:00 a.m. "Howie, don't you think you're crazy to start for Belleville on a day like this?" Mother said. "We are sure to get heavy snow by the look of those clouds."

I thought she was right but did not want to change my plans, so I replied, "I know it looks like snow, Ma, but the roads are open and old Ned will get me there and back without any trouble. Colonel Ponton will be expecting to see me and I don't want to make a new date. Mrs. Pattee has been very good to arrange for me to see him today."

So it was that after porridge, toast, and coffee, I donned the snappy overcoat I had bought a few months before, went across the field to the barn, harnessed Ned, complete with the lovely chiming sleigh bells, hooked him into the shaft of the cutter, and started on the trip to Belleville, twelve miles away. In my pocket I had a letter of introduction to Colonel W.N. Ponton, KC, a lawyer in Belleville with whom Mrs. Pattee, my old school teacher and friend, was well acquainted. She knew that I would never be happy until I got into the army and told my mother that if I saw Colonel Ponton he might get me into a non-combatant role where I would be "doing my bit" but would be unlikely to get hurt. On this basis, Mother agreed to my enlistment. Neither she nor Mrs. Pattee realized that, once you signed up, you went where you were sent. Nevertheless, it must have been a heart-rending moment when Mother heard the bells and saw me – her fair-haired boy of seventeen years – going down the road alone, the third member of the family to leave hearth and home within six months. My oldest brother had married, my father had died, and now I was going she knew not where.

Once old Ned and I got to the main road, it was good going so that we arrived in Belleville about noon. I put Ned in the shed at the City Hotel, rubbed him down, put on his blanket, fed him his oats, and then went myself to Dickens' Restaurant and Bakery and had a good bowl of oyster stew. Twenty-five cents, I think it cost. Then, just around the corner, I went to Colonel Ponton's office. He was very kind. He understood my mother's wishes and said, "Well, Graham, I think I can help you. I know Captain Doyle, who is a quartermaster of the 155th Battalion, and he may have a place for someone with your good education in the battalion stores." I thanked him and said that would be fine. He phoned Captain Doyle there and then and said he had a young man in his office four months short of being eighteen years old, with junior matriculation, who wanted to enlist but his mother wanted him in a safe job. Could Captain Doyle use him in his stores? A few more exchanges and he hung up, turned to me, and said, "Captain Doyle will be glad to have you, so go down to the recruiting office and they will arrange for a medical examination and sign you up if you're okay."

I thanked him warmly and did as I was told. At the recruiting office, I was sent across to the Armoury for the medical examination. The medical officer was not in, but his orderly, a corporal, was there, and we had a chat while waiting for the doctor. I had been short-sighted since I was a child and always worried about the possibility that I would be rejected for the Army because of this. Now in front of me was the chart used for eye-testing. Quite easily, while waiting for the medical officer, I memorized most of the lines on the chart. When I was examined, I was found to be sound in wind and limb and had no problem with the eyes! My height: 5 feet 10 inches; my weight: 137 pounds. Not robust, but adequate; after all, I was not joining the Guards. (One year later, in England, I was 6 feet and 160 pounds. The Army life agreed with me.)

Then it was back to the recruiting office with my papers for the final "swearing in." Hurray! I was on the payroll at a dollar a day, plus ten cents "field allowance." I never did learn what that meant, but I was glad to get the ten cents. I was given seven days' leave to go home and "wind up my affairs." By this time it was about mid-afternoon and snowing large, wet flakes. I was told to go to the old Catherine Street School to get my kit and report to Captain Doyle, who would be there. The school being on my route back to Trenton, I went to the City Hotel, hitched Ned to the cutter, paid a quarter for his shelter, and drove up to the old school to get outfitted: two pairs of heavy socks, two suits, long underwear (separate shirts and drawers), housewife, holdall, cap, great coat, uniform jacket and trousers, and puttees. (The "housewife" was a small folding cloth envelope containing pins, needles, yarns, and thread. The "hold-all" was a larger container that rolled up and was tied together. In it would be a razor, toothbrush, soap, etc., and the little "housewife"!) All went in a kit bag except for the great coat and boots. I met members of the team with whom I would be working and training for the next several months, including Captain Doyle, a druggist by profession and a kind and understanding gentleman.

By the time I left the old school, it was snowing heavily and almost four o'clock. Darkness set in early; it was very quiet except for the chiming of a few sleigh bells, and Ned and I met very few other travellers. But the going was heavy, and by the time we got home it was past seven o'clock. I unharnessed Ned, rubbed him down thoroughly, put on his blanket, fed and watered him, dried off the harness, and took myself across the field to the house. I did not take my kit; I left everything in the barn because I thought the appearance of the soldier's garb would cause my mother pain and grief. Mother had kept supper for me, and for the hour or two remaining before bedtime I gave her and Ernest a detailed account of the day's happenings. Mother seemed content with the fact that I would be in the Quartermaster's Stores. Ern, I think, was proud of the fact that I had enlisted, because he certainly would have done so except for a lame leg, which was four inches shorter than the other, the result of a broken hip when he was very young.

Next morning, back at the office, Mr. Hicks, the manager, and other members of the staff made me feel good by congratulating me on joining up. We had a little farewell party on Saturday, the final day of my employment. On Sunday, I went to morning church as usual with the rest of the family and to Sunday school in the afternoon, and to church again in the evening. That was the first time I wore the uniform, freshly pressed and with buttons and badges shining. Monday was the day of departure for duty in Belleville. In uniform and with kit bag containing the spare clothing over my shoulder, I set off after a farewell wave to my mother, sister Edith, and baby brother Bruce to walk the mile or more in the tracks made by sleighs on the snow-covered road to the railway station. On the way I was met by a school chum, Ted Hogle, who, though he was a little older than I and desperately wanted to enlist, could not do so because he was an only child. His mother had recently died and his father was ill, so Ted had to stay home and look after him. However, my good chum determined to go with me to Belleville, just to keep me company, and I was glad to have him. To tell the truth, I was feeling a bit forlorn now that the time had come for me to break the ties with the family I loved so dearly.

On the way to the station, we saw brother Claude, with one of our teams, unloading a sleighload of stones for a pier to be part of a new bridge that would span the Trent in place of the old covered bridge that had served for almost a hundred years. He was on the ice below us, and we waved and said our good-byes. Farther along we stopped in at Shurrie's Drug Store to say good-bye to brother Ern. He had left our house before I was downstairs that morning. During my years overseas, Ern was to be the head and main support of our mother and the children. I say "main support," because I was able to make some contribution by assigning half my pay to Mother. This left me with little to spend, but I needed little. I neither smoked nor drank, never wore civilian clothes from that day on, and so was always able to have a credit balance with the paymaster to take care of any extra expense I would have when "on leave." Finally, Ted and I got to the station. I took the train to Belleville and reported to the regimental quartermaster sergeant about noon. At the "Q" stores in the old Catherine Street school, I was lucky to get a good boarding place within a block or two of my work, with a bedroom to myself and three good meals a day for an amount equal to my "subsistence" pay, which was in addition to the one dollar and ten cents and which would be discontinued when we moved to camp.

I often think that the quartermaster does not get the credit to which he is entitled because, though he may not spend time in the front line or hear the whistle of bullets or the shriek of shells, he has a great responsibility in providing the many services necessary to maintain the well-being and indeed the morale of the troops. My Captain Doyle was one of the best, a first-class administrator and a man concerned about the services provided to the troops, whether in the cook house, the armourer's shop, the tailor shop, the washing and bathing facilities, recreation and entertainment, or other areas. As an instance of his concern, I remember the time I was laid low with a heavy cold and a high temperature and was kept in my bed by good Mrs. Barrager, my landlady, for a few days. Who would appear at my bedside each morning? Captain Doyle, asking how I felt and if there was anything I needed. I might have been his own son.

My work involved a form of bookkeeping. Every item, whether a razor or a rifle, had to be accounted for by the signature of a recipient. When a recruit arrived at the stores, as I had done, he signed a receipt for what he had received. Periodically, an inventory of articles in stock was taken and checked against the number we had received from Ordinance Headquarters and the receipts signed by the recruits. If there was a shortage, the battalion had to pay, and this indicated a laxity on the part of the QM and his staff. The same sort of check was made with regard to rations. A daily "parade slate," filed with Headquarters in Kingston, showed how many men had been away on leave, or on a course, or in hospital, and the net number of man days for which rations should have been drawn from the Army Service Corps. If a unit had "overdrawn," the CO again was responsible, but the QM and his staff had caused the error by careless or inefficient bookkeeping and indenting – ordering rations. I was a part of the team that did this recording and ordering. It was interesting, and we had a good report. I cannot recall the CO having to dip into regimental funds to pay for shortages.

A few days after I reported for duty, Captain Doyle asked me if I could sing or knew anything about music. I said I had taken some piano lessons but had never been a singer except in church. Well, he said, the battalion was organizing a minstrel show and I might come to the Armoury that night for a tryout. I was happy to do this, and on arrival at the Armoury found that Joe Doyle, a brother of my captain, had been a producer of musicals in New York and was now organizing a two-part

53

show. The first half would be a standard type of minstrel show and the second half a caricature of the Gilbert and Sullivan operetta *H.M.S. Pinafore*. There were about sixty officers, NCOs, and men (like myself) present, and somehow or other I got accepted as one of the "chorus." Then followed several nights and weeks of learning and rehearsing until we were ready to open in the Belleville Opera House. Joe Doyle did an excellent job; he had a fine bass voice himself and, with three others, interspersed the "joking" in the minstrel part of the evening with renditions by a splendid male quartet. *H.M.S. Pinafore* went off very well, with Joe Doyle taking the part of the Captain.

This was an all-male effort, so a man took the part of the maid Buttercup. He was dressed appropriately in skirt and blouse with an ample "bustle" made of rolled newspapers to exaggerate his rear. While he was singing solo "I'm called little Buttercup, sweet little Buttercup" and prancing around the stage, the strings tying the newspapers to his rear began breaking and, to the hilarious enjoyment of the audience, he left a path of strewn paper as he performed. This was the first performance, and never again did the paper get away, which I thought was a pity because it added much fun to his already funny capers. After two nights in Belleville, we went by train for a one-night stand at the Armoury of Picton, another in Napanee, and one in Madoc. Unfortunately, the show, which was really good entertainment, never played in Trenton because there was no armoury there and Mr. Waller, who owned the only theatre, and the Battalion Committee could not agree on terms for rental.

These activities, coupled with my normal duties, going to church parade, muster parades, occasional picture shows (silent, of course), leave for an odd weekend at home, skating, and hockey matches, left few idle moments from March 13 to May 24. On that date, the 155th Battalion moved by train from Belleville to Barriefield Camp, just east of Kingston, Ontario. We in the stores had a busy week prior to that, packing our stock-in-trade for the move. An advance party of some sixty men had gone to the camp a week before to lay out the tent lines and draw from the Ordinance Depot at Kingston the various tents. The 155th Battalion, and the 154th and 156th, all from eastern Ontario, were at full strength of about one thousand each. My cadet service had taught me the basics of summer training and, apart from doing rifle shooting, "physical jerks" before breakfast, and an occasional eighteen- or twenty-mile route march, I spent my time at the job in QM Stores and got my first promotion from private to lance corporal (without pay). I think I was as proud of that one stripe as any mark or rank I may have gained in my whole career. Almost fifty years later, when staff at Army HQ reviewed my service records in order to compute my liability for pension contribution, they were surprised to see that their chief had served in every rank of the service from private to lieutenant-general – except that of warranty officer. I was never quite good enough for that very important post!

Two experiences of those days in the tented camp in Barriefield remain nostalgic memories. One was the music of our excellent brass band under Band Master Reg Hinchey, which led the battalion on route marches and provided many evening concerts in the big mess tent. The other was the bugle band. We had no keyed instruments or trumpets, as the cavalry and gunners used, but old-fashioned, short brass instruments descended from the hunting horn. First sounds heard in the early morning air were the buglers sounding reveille, followed at once by the brass band marching through the tent lines and playing lively tunes like "Marching through Georgia," "Colonel Bogey," or "The British Grenadiers" to ensure that everyone was awake. From sun-up or before until "lights out" at 10:15 p.m., we heard the buglers blow a great variety of signals, always loud and clear, preceded by the battalion's code of three or four notes, so that there would be no confusion with other units. There were the three separate mess calls for officers, for sergeants, and for men; the "On Parade" call (fall in "A," fall in "B," fall in every company!); and the defaulters' call, which required those serving a minor sentence and so were confined to the battalion lines to report on the double to the sergeant of the guard. There was the pay parade call, the sick parade call, the fire call, and at the end of the day "The Retreat," the "First Post," the "Last Post," and finally "Lights Out." Nowadays, if bugle calls are heard in army camps – and I do not think they are – they would be from a tape recorder blasting over a loudspeaker!

Apart from concerts and ball games, the big weekly event on summer evenings was going to Kingston on Saturday night to see a picture show and perhaps to eat out for a change at an excellent Chinese restaurant, The Grand Café. But it was a long, hot, and dusty walk from camp to city. The road (now No. 2 highway) was not paved and, though there were not a great many cars on the road, there were enough to raise an ever-present cloud of dust as we trudged the two miles along the roadside, until we got to the city. And so the summer months slipped by, not too difficult, not too exciting, until, in late September, we received notice to prepare for movement overseas. Medical inspections, embarkation leave, kit inspections, turning surplus stores back to Ordinance, and the many other activities entailed in moving a unit of a thousand men to the port of embarkation, Halifax, kept us busy.

One lovely Saturday evening in October, I was detailed to act as guard in one of the many baggage cars. The train stood on the siding in Kingston, when suddenly appeared my dear mother. The railway had run a special train from Port Hope, Cobourg, Trenton, and other towns in that area from which most of the men came so that relatives could come to Kingston and see us off. You will appreciate that security of information as to the exact time of our departure from Kingston was non-existent. The surprise arrival of my mother and hundreds of others was a much better kept secret. She and other ladies from Trenton had prepared two great laundry hampers of sandwiches as a treat for the boys on their way to the coast, and these I took charge of in the baggage car and never touched one until Captain Doyle, my boss, authorized their issue to the troops when we were well on our way to Halifax! The trip took about three days, with a couple of hours' stop at Moncton to do a short march for a bit of exercise and fresh air. On arrival at dockside in Halifax, we found that our transport was to be the s.s. *Northland* of the Red Star Line.

I was one of four detailed to guard the officers' baggage. The troops came aboard and, in order of arrival, were assigned first to the lowest deck and then to succeeding decks until all had either bunks or hammocks. It was a very crowded ship, indeed, when it came our turn, as guards, to be assigned quarters. The only space left was a four-bunk, inside cabin with mattresses, sheets, blankets, a mirror, and a wash basin with tap and running water. How lucky could we get! I had visions of a wonderful voyage to Britain, for we knew that was our destination. In early morning we set sail in a convoy of about ten transports with, as far as we could discern, an escort of two or three warships. Almost as soon as we were out of sight of land, a gale started to blow, and the good ship *Northland* began to pitch and roll. Young land-lover Graham's stomach seemed to turn inside out! Never had I been so sick. For two or three days I did not leave my bunk. A fellow guard, being a storesman and thus having access to the hold where supplies were kept, discovered a barrel of apples and brought a supply to our cabin. Apples I could eat and keep down, and with the help of this treat my innards became accustomed to the continued tossing and rolling of the ship, so that I was able to take my place on the dining deck. One night while we were at supper, it was so rough that a wave broke over the promenade deck, burst open the doors at the top of the stairs leading down to the dining saloon, and a great rush of water cascaded down and scared our appetites away. These violent storms, I learned, were not unusual at that time of year, and were known as the equinoxial gales. Toward the end of the voyage, I was well recovered and rather enjoyed a bit of rough weather. But by and large, the voyage I had looked forward to was a disappointment. On the tenth day we steamed into the Mersey and docked at Liverpool.

It was a dull and murky late afternoon when we left the *Northland* and boarded the train that would take us during the night to our destination at Witley Camp in the south of England. My first impression of England, when we disembarked, was the sight of innumerable chimney pots, four or five to a chimney, and most of them throwing up wisps of smoke to add to the gloom of a rainy English evening. The train, of course, was a novelty, with very light coaches divided into compartments, each holding eight or ten people, compared with our Canadian trains made up of heavy coaches with entrances only at each end. For our evening meal we were issued loaves of bread and tins of corned "bully" beef, which we consumed in our compartments with the aid of the clasp knives and utensils always carried in our knapsacks.

On arrival at Witley Camp early next morning, we found it to be "hutted," not tented, as had been the case in Canada. The march to our "lines," as the battalion area was called, was not long, and soon we were lined up and issued with two blankets each, a set of bed boards, which we would place upon two trestles about six inches high, and a palliasse, which we filled with fresh dry straw from a nearby pile. We were assigned to our huts, each of which accommodated a platoon of about thirty men. A few weeks later we moved from Witley to another hutted camp about eight miles away, Bramshott, where we spent Christmas and greeted the New Year, 1917. Shortly thereafter we were told that we were to be broken up as a unit and the personnel dispersed, some to France as reinforcements, some to holding units in the south of England to continue training and be ready for dispatch to France as needed, and some to other units in England that were part of the 5th Canadian Infantry Division. This division was to be sent to France to join the four Canadian divisions already there, which were known as the Canadian Corps. This was a sad time for all of us. When any unit – and especially one's infantry unit – is disbanded, it is like the dissolution of a family. Friendships formed by working together, living together, sharing good times and bad times, the development of a great pride and intense loyalty to a battalion all are lost. A new start must be made in strange surroundings and sometimes, as was to be my destiny, in very strange surroundings indeed.

When only a fragment of the 155th Battalion was left and I continued to wonder what my posting would be, I was told that I had been promoted to the rank of lance sergeant (i.e., I would draw corporal's pay but wear three stripes instead of the corporal's two) and that I was to be transferred to the 150th Battalion, a unit in the 5th Division stationed in Witley Camp, along with our transport officer, Captain Dick Harder, and our regimental quartermaster sergeant named Waring. The three of us were told that the 150th Battalion, a French-Canadian unit from Montreal, had been having great difficulty in dealing with the British Ordinance people, and the commanding officer of the 150th was being held responsible for a whopping big bill because the QM and his staff had been overdrawing supplies, for both men and animals. All vehicles were horse-drawn and, in addition, there were in our infantry battalion a considerable number of saddle horses for the senior officers. I am quite sure that much of the trouble with the Ordinance Corps was because of language difficulties. In any event, in due course I arrived at my new unit. It was a dismal rainy evening in February, almost dark in mid-afternoon, and after meeting the unit quartermaster, Captain F.J. Boisvert from Rivière-du-Loup, Waring and I went down to our sleeping quarters and then to the Sergeants' Mess for supper.

I think, for the first time, I was terribly homesick. Something had gone wrong with the generator that supplied power for the lights in the mess hut, and the only light came from candles plugged into empty beer bottles. I did not smoke or drink, but I think everyone else did. One could scarcely see across the table. I could not understand a word of the babble going on around me, for in spite of my matriculation in French Grammar and Literature, my ear was not tuned to the rapid fire of conversation. I was a young man, now eighteen years old and self-reliant, but this strange environment was hard to take after the recent separation from friends and comrades of my old battalion. But that was the first night, and I am happy to say that the next eight months were as happy as any I ever spent in the army. Captain Boisvert was kindness itself and spoke quite good English. In many ways his attitude toward me was as Captain Doyle's had been, that of a father or an older brother. The two store men taught me many things, including how to play stud poker and, more important, the art of speaking French fluently enough to get along famously with my French-Canadian brothers-in-arms. Being well grounded in vocabulary and grammar, I was able to accomplish this in only a few weeks. Waring and Harder never did speak or understand French and, as a result, were not too happy or too congenial in their appointments. It was left to me to prepare the indents for supplies, and I am happy to say that we kept properly within our rights and requirements, much to the satisfaction of the commanding officer, Lieutenant-Colonel Hercule Barre. On one occasion I was sent to the camp hospital for two weeks with a bad attack of boils. During this time, Waring and Boisvert did their best with the indents but, alas, at the end of the month, the co got a sizeable bill for "overdrawn" rations and there was hell to pay!

In addition to work in the stores, I took my fair share of training – route marches, rifle and machine-gun firing, bayonet fighting, etc. It was during an afternoon of this bloodthirsty drill that I had my first glimpse of royalty. George v and Queen Mary visited the 5th Division, and the King with his retinue of senior officers walked through the training area, stopping to say a word or two to some of the men at their various training areas. When he came to our bayonet-fighting exercise, we put on a good show of thrusting at the straw bags representing the villainous enemy. The King seemed satisfied that we were bloodthirsty enough for battle and certainly full of vim and vigour. Queen Mary and her lady-in-waiting were in the very high Rolls-Royce that moved slowly along the road as His Majesty proceeded on his tour of inspection. Little did I think, on that sunny day in Witley Camp when I first saw my sovereign, that in nearly twenty-three years I would meet his son, King George vi, and conduct him through the area occupied by my unit, the Hastings and Prince Edward Regiment, in Aldershot Camp, and later lunch with him and other senior officers of the 1st Canadian Division.

After the first few weeks I felt perfectly at home with the 150th Battalion, made many close friends, and developed an affection for my fellow French Canadians that has not faded over the years. I can understand and sympathize with so many of them who feel that through no fault of their own or of their ancestors they have been "lorded over" by an Anglo-Saxon element and influence. But many French Canadians see only one side of the coin and fail to acknowledge the generous treatment they have received, in many ways, under Confederation.

I had enjoyable leaves of absence during my months in England. On arrival, we had a short disembarkation leave to London. On my first trip to that great city, about which I had read so much, I stayed at a leave centre operated by a voluntary group of ladies on Charing Cross Road, just north of Trafalgar Square – a shilling a night with free breakfast. Each day char-a-banc tours took us to points of interest in and around London, everything from Buckingham Palace, Hampton Court, Windsor Castle, the Tower, the Abbey, St. Paul's, even Covent Garden, Petticoat Lane, and Smithfield Meat Market. In the evening the shows were many and varied, within a few minutes' walk of my digs and very cheaply priced for the service boys. Needless to say, I was propositioned on countless occasions by attractive young ladies, but being a good church man and well warned about the danger of VD, I resisted all proposals and at war's end returned home still a virgin!

After this first visit to London I made a few day-long leaves there. My later leaves of a week or more I took in Scotland, where some of my kinfolk were still living in Edinburgh and Dunfermline, the one-time capital of Scotland where I visited cousins and visited the old Abbey Church where was buried the heart of Robert the Bruce. I also travelled to Glasgow and so saw the Highlands and the Lowlands and relived the oft-read poems and books of Walter Scott.

The Canadian Corps suffered enormous casualties in 1916 in what Field Marshal Haig called "a policy of attrition" during the Battle of the Somme, which commenced July 1, 1916, and went on for months. The theory was that by keeping the pressure at great heat and continual attack, the Allies would drain the German army to a degree where they would eventually run short of manpower, equipment, and supplies and so have to throw in the sponge. But this was not to be; it was the Allies who fell victim of the policy and finally had to call a halt after practically no gain as far as strategic positions were concerned. The Allies and particularly the Canadians suffered heavily the next year at Vimy Ridge and Passchendaele, and reinforcements were not forthcoming from Canada to keep the four divisions of the Canadian Corps in France up to battle strength. Conscription of men in Canada produced some men but not enough. The result was a change in plan for the 5th Division, of which the 14th Infantry Brigade was a part. Instead of going to France and serving as a formation in the Canadian Corps, the infantry element was broken up and dispersed as reinforcements to units in France or sent to holding units in England for dispatch to France as required. The 5th Division Artillery, however, was kept as a formation, and eventually, in

1918, moved to France and served a most useful purpose as a reserve of artillery for use by Lieutenant-General Arthur Currie, the Canadian Corps commander.

These changes meant that I had to remove three stripes and a crown from my sleeve and revert to my permanent rank of private with commensurate pay. The 14th Brigade Headquarters being disbanded, I went to a holding unit and took refresher courses on various weapons – grenades, trench mortars, rifle, and Lewis machine guns. One day I was on a range, firing rifle grenades, when a runner appeared and said I was wanted at the orderly room at once. What now? The adjutant informed me that I was on a draft for the 15th Battalion in France and would leave next morning for Folkestone with a group of reinforcements. Next night we stayed in an empty factory at Folkestone and the following morning boarded a troop ship for Boulogne, France. On deck were large boxes filled with life preservers and we were ordered to wear them. I got mine and in a few minutes was covered with lice. The preservers were crawling with the little beasties, and from then on I was seldom free of them. We disembarked from the cross-Channel steamer and travelled by railway in those little French boxcars to the city of St. Pol, which was then the end of the line, the "railhead." Here we transferred to a narrow-gauge railway and spent the night in a barn built in a square around a very stinky wet barnyard. Three or four inches of snow had fallen. We supped on our iron rations, bully beef and hard tack. With the horses, goats, and cattle in their stalls below us, we had a quiet comfortable night in the hay loft.

Next day, after a long tramp in heavy marching order, we were led to our unit. Why I was sent to the 15th Battalion I will never know. The 15th was from Toronto, whereas I was from eastern Ontario and would assume that I would go to an eastern Ontario battalion like the 2nd, the 21st, or the 39th. However, I soon learned that I was in a fine unit. I suppose the fifty or so men were desperately needed as reinforcements, so the personnel people in Britain sent whoever was available.

The story of trench warfare in World War 1 has been told so many times that I am not going to dwell on it here. For the most part it was monotony, discomfort, occasional bursts of activity, ever-dangerous enemy raids, or raids on our part to try to capture one or more of the enemy in the hope of gaining information and strengthening our dugouts in the front line to escape the effects of enemy gunfire. The Germans were doing the same. It was during one of these "stonks," as we called them, that I got a small piece of shrapnel in my backside. It was not serious and I thought I might be patched up at the regimental air post or the field dressing station, but I ended up farther back, at the casualty clearing station. Here, after a couple of weeks, I was declared fit for duty, and I assumed that I would be returning to my battalion, the 15th, a unit of the 48th Highlands of Canada, one of our greatest Militia units. I had spent only two months with the 15th, but further service with that battalion was not to be. Perhaps, from my records, someone had discovered that I had been a brigade orderly, room sergeant, and that I spoke French and therefore I would be more useful at headquarters than I would be in a battalion. Whatever the reason, I was sent to Canadian Corps Headquarters, then at Camblain L'Abbe, a hamlet west of Arras. On arrival I was assigned as a senior clerk to the Artillery Branch. My responsibility was the receipt and dispatch and recording (not on tape, but in a book) of all messages, operation orders, and instructions. I found the work intensely interesting because it brought me in touch with not only run-of-the-mill affairs but also "top" and "most" secret plans for future operations. From the end of March, 1918, these operations were many and varied and led to the final and complete defeat of the enemy forces and the entry of troops of the Canadian Corps into Mons on the morning of November 11, 1918.

There is no need to go into great detail about the events of these seven months; suffice it to say that at the Corps Headquarters we were kept extremely busy and moved about twelve times in six months. In those days, a Corps Headquarters (divisional and brigade) was relatively small. The Artillery Section of Corps could pack up and move on three or four hours' notice and with the use of only two or three lorries. For a time we had the Prince of Wales attached to the Canadian Corps. He was shown on the nominal roll as "Capt. H.R.H. The Prince of Wales, of the Grenadier Guards," his equerry or aide was "Major The Lord Claude Hamilton, of the Scots Grays." He walked with a marked limp, having been severely wounded in the first few days of the war while serving with his famous cavalry regiment.

On one of our moves we took over the building occupied by a British Corps, and one of the "limey" lads said to me, "I hear the Prince of Wales is with you people now."

"Yes," I said. "He's been with us a few weeks."

"Well, mate," he replied, "keep an eye on 'im, 'e prop'ly bitched up things when he was with us."

I didn't have time to go into the details of how he "bitched up things," but did hear he was something of a nuisance because he wanted to go right forward where the bullets whistled. The Corps Commander, I suppose, had a responsibility for the Prince's safety when he was with us. With Lord Claude Hamilton's help, he kept him within bounds, so to speak. He left us before the Battle of Amiens, and the next time I saw him was in 1920 when he opened the Canadian National Exhibition in Toronto.

One fact I learned from the top-secret files that I have never heard or seen mentioned had to do with the Canadian Corps organization. Apparently Sam Hughes, the Minister of Militia and Defence, had intended to recruit six infantry divisions for service on the Western Front, with the necessary artillery, engineers, and supply sources to form two Canadian Army Corps under a Canadian Army HQ. Lieutenant-General Currie strongly opposed this plan because it would be impossible for Canada to maintain, at battle strength, a force of this magnitude. Fortunately his advice was heeded, with the result that, despite the heavy casualties in the Somme offensive in 1916 and in the battles of Vimy Ridge and Passchendaele and in intervening operations, the Canadian Corps was able to maintain its organization with four divisions, each with four battalions to a brigade and three brigades to a division with supporting arms and services, at full or nearly full establishment right to the end of the hostilities. The infantry units of the 5th Canadian Division in England had to be broken up, as I have said, to reinforce the four divisions in France. Even so, it was necessary to introduce conscription in Canada in 1917, causing a political upheaval, in Quebec in particular, that threatened the very life of Confederation. British and French formations, unlike the Canadian Corps, had to reduce their divisional organization from four to three battalions to a brigade.

Late in July, it was known to a very few senior officers and substaff (of which I was one) at Canadian Corps that Marshall Foch, who was now the Supreme Commander of Allied Forces on the Western Front, planned a counterstroke against the enemy south of the Arras front and in front of Amiens to restore the front as it had been before the German offensive in the early spring of 1918. This planned attack, which would be launched in early August by the 4th Army under General Sir Henry Rawlinson, would have the Canadian Corps in the centre, with Australian and British Corps on our left and the French on our right. This plan involved moving the Canadians from the 1st Army on the Arras front to the 4th Army on the Amiens front, some forty miles to the south. Most elaborate precautions were taken to mislead the enemy as to our destination and our plans. Part of our 2nd Division was moved north and attached to a British formation in the front line, where they would be identified by the enemy who would assume that the Canadian Corps was in that area. The Corps was withdrawn from in front of Arras, and it was thought by some that we were going to 4th Army Reserve for rest and refit after many months of continuous contact with the enemy. By roundabout moves, we did get to an area behind Amiens. Then in a matter of days, moving only in the dark of night, the Corps, with supporting arms including tanks (which I then saw for the first time), moved into assembly areas close up to the British units in the front line. All our guns had been calibrated, i.e., checked for accuracy, so that when the battle opened they could fire over open sights into the enemy forward positions and then lift in successive lines ahead of our advancing troops. It was rumoured that when General Currie was asked for the date that he could have his Corps in the assembly areas in front of Amiens and ready to attack, he replied "by August 10th." He was told it must be done by the 8th, and it was.

I recall being on duty one night when a dispatch rider appeared with a message from an artillery commander saying he would not be able to reach his appointed place of "hiding" (it was a small woods) before daybreak. Should he continue for perhaps an hour after dawn or should he pull off the road and try to camouflage his battery during the daylight hours? I wakened the duty officer, Colonel Crerar - later General H.D.G. Crerar, Commander of the Canadian Army in World War ii - and asked for instructions. He said, "Tell him to go on to the woods, even if he has to use an hour of daylight." But this was an exception.

The entire Corps, by use of trains, lorries, and marching, moved for more than a week entirely by night, sleeping or trying to sleep when well hidden during the daylight hours. The planning, the allotment of roads, the provision of transport, trains, and trucks to move not only personnel and guns but also to establish dumps of ammunition for the projected battle, the establishment of hospitals, casualty clearing stations, and other medical facilities, all these and many other requirements to bring the Corps to its "jumping off" line about eight miles east of Amiens in time for zero hour was a very complicated and arduous task. Bearing in mind that the Corps had a strength of more than 100,000 men, plus horses and thousands of tons of supplies and equipment (probably more than Wellington had in his whole army at Waterloo), the magnitude and speed with which the shift from Arras to Amiens was successfully carried out, mostly at night, is difficult to imagine. The Battle of Amiens began at 0430 on August 8, 1918. It ended for the Canadians, after an advance of up to fourteen miles and occupation of an area of sixty-seven square miles containing twenty-seven towns and villages, on August 20.

If I have dwelt a little on the Battle of Amiens, it is because Canadians have never appreciated its vital importance as the first step in the final defeat of the German armies and the fact that it was a distinctly Canadian operation. Canadians have heard much of the Somme, Vimy Ridge, and Passchendaele. The Battle of Amiens deserves a place in our history and our memory that is second to none of these. The author of the account of the Amiens operations in *The Times* of London noted: "In structure, it was chiefly a Canadian battle . . . it was their advance that was core and crux of the operation and on their progress depended the advance of the Australians on the Canadians' left and that of the French on the Canadian right. . . . The Canadians, I think, are right, in claiming that the fighting of these first two days was the biggest thing Canada has done in the war, not excepting the capture of Vimy Ridge. Certainly nothing could have been better." Even the German general, Ludendorff, wrote later: "The defeat of our arms on August 8 . . . resulted in our losing hope for a military victory. The cost to the Canadian Corps in casualties during the Amiens operation was heavy, but not unduly so, considering the progress made and the number of German Divisions completely decimated." Canadian losses were 1,814 killed, 9,103 wounded, and 495 missing.

From the Amiens front, the Corps moved north again by a shorter route and under more favourable circumstances than the move to the south. There was not the need for great secrecy, and most of the transfer took place in the hours of daylight. So it was that the Canadians were able to launch their first great attack against the Drocourt-Queant switch, an important part of the famed Hindenburg defence line, on Monday, August 26. The depth and strength in men, mines, weapons, and wire of this Western German "Wall" was tremendous. I recall looking across a valley on a sunny afternoon and noting what appeared to be fields of ripened buckwheat as far as one could see. The buckwheat was roll upon roll of rusted barbed wire. Through this formidable obstacle, our gunners had to blast lanes for the infantry to pass. Our attack on the Drocourt-Queant switch was successful, as were succeeding assaults, all timed and made in conjunction with British, French, and Australian formations. From August 26 to November 11, our casualties in killed, wounded, and missing almost reached 40,000.

The personnel in the Artillery Branch at Corps was notable. The GOC was Major-General Sir Edward Norrison. We called him "Dinky" for some reason. His stepson Herbert Fripp was a staff captain and was entitled to and did wear a red band on his hat, red tabs on his collar, a red-and-white four-inch band on his sleeve (corps colours), and reddish Sam Browne belt and riding boots. He was very short and we disrespectfully referred to him as "the little man dressed in red." One day Herbie told a senior artillery commander, "Father wants you to do so and so." As the story goes, the officer replied, "Tell your Pa to go to hell, I only take orders from the General." But Herbie was a hard-working and efficient young man.

The counter battery staff officer was Colonel Alan Brooks, later to be the Chief of the Imperial General Staff and Churchill's prop and stay in World War II. He was succeeded by Colonel McNaughton, later to be Canadian Army Commander in that war. He, in turn, was succeeded at Corps Headquarters by Colonel Crerar. The hand of fate brought Private Graham, when he had become a senior officer, in contact with some of these officers a quarter-century later.

Among the substaff of about twenty-five, including signallers, dispatch riders, stenos, orderlies, and filing, recording, and dispatch clerks, was one Tom Gentles from Halifax. Tommy could take dictation and rattle a typewriter as well as any court reporter. His hair was as red as fire and his face a mass of freckles, so he was called "Red." Most of us had nicknames. Mine was "Burt." Apparently I resembled one of the characters made famous by Bruce Bairnsfather, the well-known cartoonist. It was customary for a soldier to team up with someone, and in this case it was Red and I. We had been in France almost a year and well past the time when we were entitled to a leave of absence. So early in November, 1918, we applied for and were granted leave for two weeks, from November 9, to go to Paris and Cannes on the Mediterranean. The tide of battle had ebbed and now it was a case of "pursuit." The pressure of work had eased up and leaves of absence were being granted.

Red and I had warrants that gave us free train passage from Corps HQ in the city of Denain to Paris, but from there to Cannes we had to pay our own way. However, we had good credit balances in our pay books and saw no problem in spending a week in Paris followed by a week on the Riviera. The only immediate problem we faced was getting from Denain to Paris, for the route we had to take was by foot, by transport, and by train through Somain, Arras, St. Pol, and Rouen to our destination. We arrived, exhausted, at St. Pol late Sunday evening, November 10, and took a room in a decent-looking estaminet. We had a good supper and decided to "hit the hay." Sometime later we were wakened by a tremendous commotion and singing from the barroom below us. We heard repeated shouts of "Hurrah, hurrah! La guerre est fini, la guerre est fini!" Later we wondered how news of the impending armistice, effective 11:00 a.m. the next morning, got to this estaminet at St. Pol. Red and I paid little attention and were too exhausted to be long disturbed by the raucous celebrations. Of course, we knew nothing of the reason for the merriment. It was only the next morning that we learned of the signing of the Armistice. Everyone was in high spirits. We did manage to get a few hours of rest and eventually we got a regular passenger train to Paris.

On arrival at the Gar du Nord, with a few dozen other soldiers on leave, we were assembled by two or three red caps (military police) and marched to a nearby hall. Here we were given a fifteen-minute lecture by a medical officer on the dire results that might come from fornication. Just in case we could not resist the importunities of the Parisiennes de la rue, each of us was given a packet of condoms and a tube of disinfectant ointment. After three nights and three days of travel, we were free to enjoy ourselves.

We checked into the Hotel d'Ienna, which had been taken over by the Canadian YMCA, and after a meal in the dining room we explored the sights and sounds of the City of Light. After four years of darkness and danger, the bright lights were being restored. There was bedlam. The Champs Elysées was thronged with people, dancing, singing, and shouting. Bands were playing. There were lines of people marching as in prison, with hands on the shoulders of the ones in front. Many had climbed the bare trees that lined the avenue. From these perches they waved flags and streamers. We were stricken dumb, but not for long. We were soon part of the frolic.

American uniforms far outnumbered the combined lot of British and French. Many of the Yanks had seen service a few months before the Armistice when they stemmed the German thrust toward Paris. But thousands of the boys from the United States who celebrated on leave in Paris during those days had not yet "smelled powder" and now, of course, they never would. "La guerre est fini!" In this happy, carefree crowd, the French civilian population took a leading part, and not least conspicuous among them were the chic, petite mademoiselles!

The following days were spent on conducted tours of Fontainebleau,

St. Cloud, Versailles, the Louvre, Malmaison, Notre-Dame, Les Invadides, the Opera, Les Follies-Bergères, and many other great sites. Since then I have been in Paris many times but I have never seen so much as I did during that week. The cost was little, and I have never failed to be grateful to the societies, staffed mainly with volunteer workers, that provided such wonderful entertainment and accommodation for the troops.

Then King George v visited Paris. Enormous crowds welcomed him, and Red and I were fortunate to be in the front row when his open landau, drawn by four horses, with the President beside him, circled the Place de L'Etoile, preceded and followed by a brilliantly uniformed cavalry escort. Notices were published that a great parade of Allied troops would assemble at the Place de L'Etoile and in the avenues that radiate from it, like rays from a star. We were told where we should assemble with other Canadians who might be in Paris if we wished to join the parade. Of course we did. In the centre of the Place is the Arc de Triomphe built by Napoleon to commemorate his victorious campaigns. After the defeat of the French by the Germans in 1870, a heavy iron chain was erected around the Arc, and neither vehicles nor pedestrians passed through the arch. Now, at last, after almost fifty years, on this glorious sunny November Sunday, the chain was cut, and with the thousands of others in that great victory parade of 1918, I marched to the sound of thrilling music under the Arc de Triomphe and down the Champs Elysées. We were to march only to the Place de la Concorde and then disperse, but this was not to be. The crowds along the route were tremendous. We were perhaps three-quarters of the way to our destination when the parade simply disintegrated in the crowds surging through us. Red and I were grabbed by two comely wenches and led off to a bench along the Seine not far away, where a little lovemaking took place. Still being good boys and the mademoiselles being nursemaids, we parted company after a pot of tea and a bun at one of the many sidewalk cafés.

The next evening we said farewell to our comfortable room and the bright lights of Paris. We boarded the *Rapide* for Marseilles and took a second train to Cannes. We reached Cannes about noon, greeted by warm sun and balmy breezes. Just across from the railway station we saw a rather attractive-looking Hotel Etrangers and across we went and registered because the rate was reasonable and the place looked clean and comfortable. Then with our base in Cannes, we saw the sights of Monte

Carlo, where we were not allowed in the gaming rooms because we were not officers and lacked evening clothes, and took the train to Nice a couple of times, even bicycling to the town of Grasse above Cannes, noted for its perfume factories.

The war was over; we decided that there was no hurry to get back to the Corps, and we might as well stay in the south as long as our funds lasted. Therefore we put aside enough for railway fare from Cannes to Paris and a few francs for food on the way. When we were down to just about our last *centieme*, we checked out of L'Etrangers, bought two loaves of crispy French bread and a tin of corned beef, and boarded the train for Marseilles and then headed back to Paris. We had already overstayed our leave period by more than a week. In Paris, we reported to the Canadian leave centre and were told that the Canadian Corps was moving through Belgium and we should go next day by train to Rouen and wait there for further orders or information. As directed, we took the train to Rouen and there we were directed to the transit camp, a large number of bell tents for transients like us and a few huts for the staff. And transients there were, in large numbers and varied nationalities: Italians, French, British, Canadians, Americans, etc. There was great, unavoidable confusion. It was cold and conditions were cramped. We were there about a week.

About December 8, the fifteen Canadians in the transit camp were sent by train to Liege in Belgium and then on to Bonn-on-the-Rhine in Germany. We had been "on the loose" for thirty-three days, and we had beaten the main part of Corps HQ to Bonn by a few hours. When they arrived by lorry, Red and I welcomed them, and they us, like long-lost brothers. The 1st and 2nd Canadian Divisions had moved into Germany and formed bridgeheads across the Rhine at Cologne and Bonn. The 3rd and 4th Divisions and most of the supporting units of the Canadian Corps were returned to Canada and demobilized during the winter and early spring of 1918-19. Our stay in Bonn in the army of occupation was short, less than seven weeks.

During that time Red and I and most of the substaff on Corps HQ were billeted in a hotel. Our offices and some officers' billets were in the attractive buildings of the University of Bonn, the grounds of which ran right down to the river. The German people, very kind by nature, showed no animosity toward us, and, I think, we reacted accordingly. Electric railways operated along the west bank, and we went frequently to shows and operas downriver at Cologne and upriver to Mainz and Wiesbaden.

The perils of war passed, the days passed quickly, work was light, and we played baseball with the Americans, who were on our right, and soccer with the British on our left. The winter was fair and warm. On January 26, 1919, we packed our gear and left the pleasant surroundings and our friendly former enemies in Bonn, and went by train to the small town of Jodoigne in Belgium, about twenty miles from Brussels. From here, Corps HQ organized and administered the withdrawal of the 1st and 2nd Divisions from Germany to staging areas in Belgium and thence to Britain. Leave was granted for two or three days to visit Brussels. Red and I found it to be a lovely city, and we explored it thoroughly by conducted tours in two days and saw shows at night. Back in Jodoigne, we did our jobs but were looking forward to the day when we might be told it was our turn to join a draft for England and return to home, sweet home. But I contracted the dreaded Spanish influenza, which was epidemic throughout most of the world at the time, and was hospitalized at the English Army Hospital in Charleroi. Those weeks are a hazy memory. When I was sufficiently recovered, I was moved by hospital train from Charleroi to a Canadian Army Hospital at Etaples. The nurses and staff at Charleroi could not have been more kind and attentive to this ill and lonely young Canadian soldier. Strangely enough, I have a vivid recollection of looking out the window of the hospital train from my upper bunk and seeing, once more and for the last time, the bleak, barren, blasted countryside of northern France and Belgium, over which the tide of battles had raged so recently. Gaunt trees were here and there, and the entire region was scarred by shell-holes, trench lines, barbed-wire entanglements, shattered buildings, all partially hidden under a recent fall of snow. It was indeed a sorrowful and depressing sight.

I was in hospital for a few days at Etaples, perhaps waiting to gain strength or until a hospital ship was available at the nearby port of Boulogne to carry us across to England. In early April, I was moved by ambulance to the dockside at Boulogne where was tied up a beautiful white hospital ship. As was done with the other bed patients, I was carried piggyback by a husky German prisoner of war from the ambulance to the top of a gently sloping, very smooth chute. He carefully placed me there and I slid down to a lower deck and was put in a bunk – white sheets, very posh – and in due course heard the engines start, and we were on our way to Dover. When we arrived at dockside in Dover, it was dark. I was carried by stretcher to the nearby railway platform, and by an English hospital train taken to No. 4 Canadian

General Hospital located in what had been an English psychiatric hospital before the war, at Basingstoke in Hampshire.

I was at Basingstoke a few weeks recuperating. Soon I was helping the Canadian nurses make the beds in my ward, and to this day I can still tuck the ends as well as my wife! One nurse who brought the beef tea (Oxo) in mid-morning said I should have a bottle of stout once a day to "build me up," and thereafter an orderly came in each morning with a glass and a bottle of good black Guinness stout on a tray for Private Graham! The stout and good care soon built me up so that I could don a walking patient's light blue trousers and jacket, white shirt, and red tie, and go into town to an afternoon show, or just for a walk in the lovely Hampshire countryside. By late April, I was discharged from No. 4 and sent for a few days to a holding unit at Seaford, on the south coast, and from there to Rhyl staging camp in North Wales, south of Liverpool, to await passage with hundreds of others on a ship to Canada and home. I was then twenty years old.

At Rhyl, which a few weeks earlier had seen rioting by Canadian men tired of waiting around to be repatriated, I had not long to wait before I was on draft for Canada from Liverpool in the s.s. *Saturnia*. This time I was not as fortunate as I had been in 1916 when I shared a second-class cabin with three of my pals. Now I drew a hammock, slung from beams in the mess deck, so that when I went "to bed" I climbed from deck to bench to table top, hooked up my woven cord hammock, and climbed in. In the morning, wakened early, I reversed the process, tumbled out of hammock onto table, thence to bench and deck! When the hammocks were stowed for the day, we washed up, showered if necessary, fell in line for our meal in mess tins, and ate it on the aforesaid tables. It really was not as bad as it may sound, because we had a lovely crossing from Liverpool to Montreal, only about six days on an open sea that behaved in a calm and gentle manner. I got my first look at great floes of ice and many icebergs as we approached the coast of Newfoundland. Mid-May was a good time for a North Atlantic crossing, some foggy mornings but generally sunny and warm days.

On May 22, 1919, in early morning, we berthed at Montreal. The troop train was at the dockside and before long we were on our way westward. Many, including myself, detrained at Kingston where we would be quickly processed and discharged, while the balance of the trainload of veterans would go on to Toronto. Now I had to make some decisions that would carry me through the next lap of my road.

5

LAW SCHOOL

Ｉt had always been my intention to continue my education and
achieve a university degree in one of the professions. But like
countless others, I had to make my own way without help from the
family. My mother, from the pay I had assigned to her during my army
service, had saved more than $300 and bought Victory Bonds. How she
did it I do not know, but with her good Scotch background she was
always a great manager. In addition to this little nest egg, I learned from
inquiry at the offices of the Federal Department of Soldiers' Civil Re-
establishment in Toronto that, having enlisted at age seventeen, I was
entitled to the top amount paid by the government in cash to assist
veterans until they could get established in civilian life. This grant, in
my case, was $480 payable at the rate of $80 per month, with a further
grant for the purchase of essential books and, as I recall, for the first year's
tuition if one went to university.

After much thought I opted to study the law. It came about this way.
My cousin-in-law, Harry Christie, a senior executive in a manufacturing
company, knowing that I had a bent for the law, arranged for me to see
his legal adviser, H.M. Mowat, a nephew of Sir Oliver, one-time Premier
of Ontario, head of the old and highly reputed legal firm of Mowat,
MacLennan, Hunter, and Parkinson. Mr. Mowat, like cousin Harry,
thought I would like and should be successful in the profession because
of my good grades at high school, and offered to take me on as an articled

student. The upshot was that, in September, 1919, I entered first year at Osgoode Hall Law School, at that time the only place in Ontario where one could prepare for the profession and become qualified as a barrister-at-law and be accepted as a solicitor of the Supreme Court of Ontario and a notary public.

During the summer, I worked at the Canadian Pacific Railway yards icing refrigerator cars and as a car checker. I was kept busy until after Labour Day. Then I went off to Toronto, where, like hundreds of other young students, mostly war veterans, I found a lodging within walking distance of Osgoode Hall on Queen Street and the Mowat & Company law offices at the corner of Richmond and Yonge streets. I shared a room on the top (third) floor of a rooming house with a pharmacy student. Almost everyone in the house was a young veteran. No. 18 Dundonald Street was but one of a great many residences in the area that accommodated hundreds of university students at very low cost. A nearby Chinese laundry did our washing. We usually pressed our own clothes. We were a congenial, happy, and temperate group. We knew the best places to eat. The Three Castles Chinese Café and its opposition, the Six Castles, both served an excellent evening dinner for thirty-five cents and were so busy from 6:00 to 7:00 p.m. that one had to keep a firm grip on one's coffee cup or the waiter (maybe part-owner) would pluck it from one's hand if one's plate was empty! In the morning, on our way to 9:00 a.m. lectures at Osgoode Hall, we usually had a hasty breakfast at Bowles Restaurant across from the City Hall. Coffee was five cents, cereal ten cents, toast and jam or baked apple ten cents. For lunch, we usually patronized one of a number of tea rooms run by elderly ladies who did their own cooking and baking. The Peacock, near our office building on Yonge Street, was a favourite, and thirty or thirty-five cents was good for a tasty home-cooked lunch.

In the fall of 1919, the old frame Arlington Hotel, at the corner of King and John streets, was still being operated by a voluntary women's organization as a sort of hostel for returned servicemen, providing bed and breakfast for a nominal sum and serving hefty midday and evening meals for a quarter. We often went there to eat until it closed at the end of the year. It was a pleasant old place, with a verandah all the way across the front that was well supplied with old-fashioned rocking chairs. The Queen's Hotel on Front Street, now the site of the Royal York, had a similar verandah where one could relax with feet on the railing. Here I stayed and sat a few times after I became an affluent young lawyer!

73

I have been referring to "we" and "our." The group consisted of four students articled to the four members of the law firm. We were Howard Green, George White, William MacRae, and myself. Howard Green, after graduation, returned to his home city of Vancouver and later became a federal member of Parliament and Minister for External Affairs in the Diefenbaker cabinet. George White, after I had left the firm to start my own practice, did likewise and opened an office in his home village of Madoc, about forty miles north of Trenton; later he, too, became a federal member of Parliament, was appointed to the Senate, and for a time was Speaker of that august body. William MacRae, who left the firm after I did, became a well-known patent lawyer in Ottawa. Our routine was to attend lectures from 9:00 to 11:00 a.m., then visit our office to learn the practical side of the profession until 4:00 p.m., then attend the Hall for one or more lectures. This pattern persisted for three years, with long summer vacations away from Osgoode and the office.

We knew little of use to our firm when we started, but in a short time we knew how to search titles in real estate transactions; how to draw deeds, mortgages, statutory declarations; how to prepare papers for the probate of wills or the appointments of administrators of estates of deceased persons; and even how to appear in division courts (now small claims court) for a plaintiff or defendant. In our firm, Mr. Mowat, a King's Counsel, did most of the litigation in the higher courts, and since I was articled to him I searched the various law reports and digests, as he sometimes required of me, to find previously tried cases in Canadian or British courts he could cite to the trial judges in support of his own arguments. This work I found intensely interesting and, in my later practice, after due research, I could write good opinions and was reasonably articulate and successful in court. But I never really liked the litigation side, which involved appearances before judge and jury or judge alone.

Mr. Mowat was extremely kind, as were all the members of the firm, though rather pompous in appearance, action, and speech. One day he said to me, "Graham, have you entered an appearance in such and such a case?" An appearance was a short document, which a student could prepare, that had to be filed within ten days of receiving a writ of action or the plaintiff might move for judgement by default.

I was able to reply, "Yes, sir, I did that yesterday."

"Good boy," he rejoined. "Never let them catch you with your pants down!"

Note that I said "sir." Since we students had recently been in the army and subject to certain rules of discipline, we always quickly got to our feet when Mr. Mowat or any other member of the firm came into the students' room. The four of us were in a small room, each with his own desk. Quite often, Mr. Mowat would invite the four of us to his home on Wellesley Street, just east of Yonge, for Sunday afternoon tea. He and his wife, a kind and gracious lady, had no family, and I think they enjoyed having the four young men in for an hour or two. Certainly we enjoyed the change from the Three Castles Chinese Café. Mrs. Mowat had a good cook and we always had a delicious high tea.

We soon learned to be useful at the office. But we were being taught by our masters at the same time. Mr. R.J. (Roderick) MacLennan was a specialist in insurance law and, though I never knew of him going into a courtroom, he was a great student and highly regarded for the opinions and briefs he prepared for counsel at trials involving insurance claims. We were trained by men who were themselves scholars devoted and dedicated to their profession. We were paid at the going rates, which were $4 per week for first-year students, $6 for second-year, and $8 for third-year. One day we decided we should have more pay! Respectful as we were, we felt the going rates were too low, especially as we were veterans and older than most articlers. We decided that Howard Green, being a little older than the rest of us, should be our spokesman and ask Mr. Hunter if we might have an interview with him. This was quickly arranged and we trouped into R.G.'s office. Affable and polite as always, R.G. said, "Sit down, boys. What can I do for you?" Howard Green then explained shortly, but clearly, that none of us had much money, what we had saved during the war was running low, we were now doing work for the firm that was producing handsome returns, and we thought we should be paid a higher salary.

R.G., I am quite sure, foresaw what was coming when Howard had asked for the appointment. After hearing our plea, he smiled and, like a good diplomat, said, "Well, Green, I appreciate what you have said, and you boys are doing good work. But you know the overhead in an office like this is very high, so that all that you make is not profit. However, I will talk to the other partners and see what we can do. Now come on down to the Board of Trade and have lunch with me." And so we did. It was a nice way to end a friendly talk. The result of our office insurrection was that we got a 50 per cent increase a week later. This meant we would receive $6, $9, and $12 a week for first-, second-, and third-year students.

This was a lot to us, when meals cost less than a dollar a day, lodging about $5.00 a week, and everything else accordingly. Certainly we were better off than most students at university, who had no income whatever except what they might earn during the long vacation.

We were able to enjoy cheap seats at many shows. We saw Harry Lauder, the great Scottish comedian, the foremost Shakespearean actors of that day, and heard visiting speakers. I recall sitting in the top gallery at Massey Hall and hearing Earl Balfour, British foreign secretary and author of the Balfour Declaration, which favoured a Jewish national homeland in Palestine. Another world-famous figure of that day, David Lloyd George, the British Prime Minister during World War I, also spoke at Massey Hall. I recall again sitting high up in the balcony, a long way from the stage where this diminutive Welshman with flowing white hair declaimed in a voice so pleasant to the ear and in perfect clarity, without distortion through the use of loudspeakers, words that went something like this: "And during those years, I stood, as it were, on the summit of a great mountain, from which I could view the vast world scene." He was referring to his days as Prime Minister and senior British representative at the Versailles Peace Conference.

Another figure, not yet notable but soon to be so in Canada at least, we heard in a West Toronto skating rink (the Ravina). He was William Lyon Mackenzie King, and among others with him on the platform was W.S. Fielding, one of the most prominent members of the Liberal Party and Minister of Finance in the last Laurier cabinet. King had but recently been elected leader of the Liberal Party in Canada and was now opening his first campaign for the office of Prime Minister. Here, again, I recall one of his main points, which was that, under a government led by him, there would be employment for everyone and "all Canadian workers would have a full dinner pail"!

As young law students, we were much interested in hearing good speakers discussing public affairs of the day. But we were also interested in hearing good preachers of the Gospel, and we regularly attended church services, usually in one of four main edifices: the Metropolitan Methodist, where the gifted Welshman, the Rev. Trevor Davies, was minister; or St. Andrews at King and Simcoe streets, where the Rev. Stewart Parker from Scotland presided; or the Baptist Church on College Street, later moved to North Yonge Street, where the Rev. W.A. Cameron preached and where they often had musical evenings with organ, string quartet, and splendid choir; but most frequently at the Rev. Dr. George

Pidgeon's Bloor Street Presbyterian Kirk, because it was convenient and also because he was a particularly lucid speaker who attracted a large congregation of university students. In 1925, he became the first Moderator of the United Church of Canada.

For the summer of 1920, George White and I moved from Dundonald Street to Centre Island, across the bay from Toronto, where we shared a room with one double bed in a cottage on the lakefront, owned and operated as a rooming house by a dear old English lady named Mrs. Spotten. Others at the cottage were a widow, Mrs. Atkinson, her grown daughter Ella, who was a secretary in the city, and a couple of other young businessmen. Mrs. Spotten gave breakfast to all of us, and then it was a pleasant walk along the quiet avenues and wooded paths of the island, where no motor vehicles were allowed, to the ferry dock each morning, and then the short voyage across the bay on one of the paddle-wheeled ferryboats to the foot of Bay Street. After our day in the city, we usually took the five-thirty or six o'clock ferry back to the island and often, after a quick dip in the lake (we were separated from the clean sandy beach by only a few yards), we could get a splendid, reasonably priced meal at the Algonquin Hotel, also on the lakefront, a block or two from our place.

Life at Mrs. Spotten's was pleasant in more ways than one, for she had a baby grand piano. Mrs. Atkinson could play and Ella had a beautiful contralto voice. Many a summer evening we were treated to a delightful collection of old songs. A friend of George White's, Harry MacKay, came up from Belleville to spend a few days as a guest. Harry had a fine baritone voice and the program he and Ella and Mrs. Atkinson at the piano put on could have done justice to a crowded concert hall. More important than the musical evenings, however, were the lessons George and I got from Mrs. Atkinson and Ella in the game of bridge. Both of us had played euchre and poker but never bridge. Our teachers were excellent players, and I think we were good pupils because we soon became adept at the game, which in my case has given me tremendous pleasure, first at auction and then at contract for three score years and more.

After our summer with Mrs. Spotten, George and I moved to board and room with Mrs. Maughan on Albany Avenue. Here again we were lucky to find a home away from home. Mrs. Maughan had three sons living at home. Ted was younger than I and a top-grade pianist who later became a teacher at the Conservatory of Music. Jack was my age and a traveller for a silverware firm. Allen was in the insurance business. We were a happy "family" because we all got along well together and Mrs. Maughan was mother not only to Ted, Jack, and Allen, but also to George and myself. Here I stayed, except for the next summer with Mrs. Spotten on the island, until I left the city in late 1922.

The lecturers of that day at Osgoode Hall were notable. Two were full-time: Dr. N.W. Hoyles, the principal, and John Falconbridge, son of Sir Glenholme Falconbridge, one-time Chief Justice of Ontario. In addition, there were four or five practising lawyers who were considered to be authorities in branches of the law. Samuel Bradford was an authority on torts; Shirley Dammison on real and personal property; and Mr. Justice W.R. Riddell on practice. Riddell had been a high school teacher before entering law and, in his lectures and judgements, he was prone to use a great many Latin maxims. In fact, Latin quotations were in much more common use by all judges than is the case today. Riddell always addressed us thus: "Good morning, fellow students" – even though he was a justice of the Court of Appeal. The point he wanted to impress upon us was that, in our chosen profession, we must always be "students of the law," which was being continually enacted or revised by governments and interpreted by binding decisions of the courts. Certainly throughout my life I have never forgotten Riddell's emphasis on the fact that there is still much to learn in whatever occupation one may be in. The law school principal, Dr. Hoyles, was a great gentleman, a wise and learned lecturer, and someone ready at all times to see and discuss with a student any problems the student might have. He recognized and appreciated that 95 per cent of the classes of 1919 and 1920 were war veterans and, in many ways, more mature than non-vet students.

I should have spent three years at Osgoode Hall before graduating, but the benchers (i.e., the governors) of the Law Society of Upper Canada, whose responsibility it was to set the standards, the curricula, and all other matters pertaining to the law school, had arranged for war veterans to take a special summer course in 1920 so that, if successful in the first year's examination, we could become qualified for entrance to the bar of Ontario in less than twenty-four months. So it was that I and other veterans who started our studies in September, 1919, finished our first year in the spring of 1920, took our second year in the summer of 1920, and finished our third and final year in the spring of 1921. It was a long and steady grind, with only a two-week break at midsummer 1920.

The class of 1921 was the largest in the history of the law school, with more than 200 graduating students. Among them were many who became leaders in the profession and in public affairs: Cyril Carson, who was our Gold Medalist; Leslie Frost, Premier of Ontario; Kelso Roberts, Attorney General; William Nickle, Provincial Secretary; O.E. (Ossie) Lennox, chairman of the Ontario Securities Commission; John Cartwright, Chief Justice of Canada; and many other appointees to the bench as magistrates, county court judges, and Supreme Court justices.

My ambition was to become a county court judge. Having passed with honours and standing fifteenth in the class, I might have made it if other events had not intervened. My certificate from the Law Society of Canada is an impressive parchment, which certified that "Howard Douglas Graham . . . was this day (19 May 1921) called to the degree of Barrister-at-law, and was admitted to practise at the Bar of His Majesty's Courts in Ontario by the Benchers of the Law Society of Upper Canada in Convocation pursuant to the rules of the Society in that behalf, and that for proficiency in his studies he was called with Honours. . . . " There was a little catch in all this high-flown rigmarole. I could practise at the bar, but I could not open my own offices until I had been granted a solicitor's certificate, and this would not be until I had served at least a full three years as an articled student. The benchers wanted to ensure that, before we embarked on the responsibility of our own law practice, we had had at least a full three years "devilling" for our masters in law firms. It was a wise rule and it ensured, as far as I was concerned, that I would stay with the Mowat firm until at least the early autumn of 1922, always learning and absorbing knowledge of the practice of the fascinating law profession.

Dr. Hoyle, at his farewell lecture, gave us much sound advice. One piece of wisdom sticks in my mind. "Now gentlemen," he said, "you are being graduated and soon will be dealing with and advising clients as to how the law may affect their problems. Remember this, you will not, after graduation, know the law, except perhaps, a very small part of it. In these few years, we have tried to teach you some law and, most important, where to find the answers to legal problems, and so do not be hasty in advising your clients. Take time to think about their problem, tell them you would like a little time to think about it, and look up some authorities on the subject. They will respect you the more for this approach than if you give them an immediate reply off the cuff." How right he was! Even though sixty years ago the laws were fewer and simpler than they are today, we had little more than two years of tuition and practical experience under articles in a law office. Indeed, from that day forward, in my own mind I remained a "student-at-law."

The summer of 1921 saw a change in the Mowat firm. Howard Green left for his native British Columbia to practise in Vancouver. Bill MacRae went to Ottawa to join his brother in a patent law practice. Herbert Mowat was shortly to be appointed a justice of the Supreme Court of Ontario. George White and I did not become partners but were hired on a salary, with the probability of being admitted as partners after we had more experience and had received our solicitors' certificates. And so matters went along for a bit more than a year. My salary was adequate for me to board with Mrs. Maughan in the winter and room with Mrs. Spotten on Centre Island in the summers of 1921 and 1922. I could enjoy a life, if not with affluence, then at least of security. But as the months went by, I was not too happy with the type of work I was being given. It was mostly conveyancing, searching titles, drawing mortgages and deeds or leases, completing documents for the probating of wills or appointment of administrators, all for the clients of my seniors. I could not see much opportunity of developing a clientele of my own. A great many of my classmates had left the city. The Mowat firm's practice changed somewhat after the departure of Herbert Mowat to the High Court. Less litigation and more estate and conveyancing was the order of the day. I liked the members, and all were very kind and helpful, but I decided that this was not enough. I was ambitious and decided that a change might be best for my future. My home town of Trenton was not overly stocked with lawyers and my family was well known there. Perhaps I would be wise to make a change, leave the security of a regular salary with prospects of increase and of becoming a partner, and open my own practice in Trenton. I consulted a friend who had practised law in Trenton and was now employed at Osgoode Hall. His advice was strong and clear, "Go to Trenton. In a year you will be making twice what you are now."

Still uncertain, I phoned Justice Mowat in his chambers at Osgoode Hall and asked if I could come to see him about a private matter. "By all means, Graham, come along," he replied. I went and told him of my thoughts and said I would appreciate his advice. It went something like this: "Um! Yes, Graham, I see your point. I know the firm will hate to see you go. You had a good record in your studies and should be very successful in Toronto, but it will take time, whereas if you go to your home town, where you and your family are known and have many friends and relatives, I predict that in a short time you will be a big toad in a little puddle, whereas in Toronto, for a long time, you will be a little toad in a big puddle!"

The upshot of all this soul-searching was that, in October, 1922, I left the firm and opened offices in Trenton, which a few years previously had been occupied by our own family lawyer, S.J. Young. The premises were owned by Wesley Saylor, whom I knew and had met on the street when I was home for the Labour Day holidays. I told him I was thinking of opening a practice in Trenton. He was most kind and enthusiastic and said he had "the very place" for me. And so it was. The rent was $15 a month and he never raised it for the next seventeen years! He gave me a considerable amount of legal business and perhaps I charged him a very fair fee to reciprocate!

George White was a bit miffed at my leaving the firm. He had married and bought a house in Toronto, and to all intents and purposes would remain there. He had assumed that I would do the same and, since we were close friends, would be a partner with him for many years to come. But a sequel to this is that, when he saw how I had prospered in Trenton in a year or two, he also withdrew from Mowat & Co. and set up a successful practice in his native village of Madoc, north of Trenton, and, as I have said, eventually became a federal member of Parliament and Speaker of the Senate.

One event that occurred in Toronto before I returned to Trenton had a profound and lasting influence on my life from that time onward. One winter day George White said to me, "Doug, how would you like to come skating at the Mutual Street Arena tonight with Irene and me? She will bring a friend along and you will make a foursome, which will be more fun than if we go alone." Irene was his fiancée and a nurse-in-training at Wellesley Hospital. I was happy to accept his invitation, and that evening I met the young woman who a few years later would become my bride. In reading biographies, I have often wondered why so little is said about "the woman behind the man." In most lives, one or sometimes more than one woman has had an important part in the success or failure of the husband's enterprises. Certainly to my dear Jean Lowe, whom I met on that blind date at the Mutual Street ice rink, I owe an immeasurable debt of love and gratitude for her loyalty, understanding, sympathy, support, cheerfulness, and all the other virtues that have helped me through more than sixty years of married happiness. And the years have not all been easy for her, as my varied activities have taken me from the home she always made so pleasant, attractive, and comfortable

81

for so many evenings, weeks, months, and even years when she would be left alone. And no matter what my positions, she had the courtesy, poise, and dignity of manner that made me always proud to have her stand beside me. With this well-deserved tribute, I now turn to a new section of the long road of my life.

6

BETWEEN THE WARS

To embark on a practice of my own in an office in Trenton, I needed furniture, a typewriter, a safe, essential textbooks, stationery, printed letterheads, and some sets of *Law Reports*. Although I would be able to use the law library in the courthouse in Belleville, it was twelve miles away and I had no car. I had saved a little from my months of work with Mowat & Co., and my brother Claude, who was already progressing in a garage business, loaned me $400. With this I bought what was needed from a second-hand store in Toronto. I had linoleum laid on the office floor, bought a pot-bellied stove, and sent out to almost all the names listed in the telephone book a very nicely printed announcement that "Howard D. Graham . . . has opened offices for a general practice of the law in the Saylor Building, Corner Dundas and Quinte Streets, in the Town of Trenton." There was no street number and no mail delivery. I lived with my mother, sister, and little brother, Bruce, and from the very beginning paid my way.

Trenton, now a city, was then a town at the junction of three counties, Hastings, Prince Edward, and Northumberland. It was about equal distance from four fair-sized villages, Consecon, Brighton, Wooler, and Frankford. From this it will be seen that a lawyer might draw clients from a fairly wide and well-settled, prosperous community. Indeed, Trenton was the centre of an excellent farming area. When I arrived in late 1922, there were but three practising the profession. T.A. O'Rourke was also the police magistrate and rather elderly. H.J. Smith, a bachelor, had recently come from New Brunswick. Gertrude Alford was charming and able, a fairly recent Osgoode graduate a few years my senior. It was not long before I had "callers," some with a little business, some just to drop in and have a look at Wes Graham's or Jessie MacPherson's boy! On Saturday afternoons I would lock the door and mop the linoleum floor, and on Monday in winter I was at the office early to start the fire, which I tried to keep going all week. Bruce, for ten cents a week, came in after school and carried the ashes out. For almost a year I did my own typing on an old second-hand Underwood, until my sister, after a course in Toronto at Shaw's Business School, came in as my secretary. She was a tower of strength because she was not only excellent at shorthand and typing but she knew so many of the people who became my clients.

To become a successful lawyer, I found it useful to become a "joiner." The Rotary Club had just been organized in Trenton and, there being a vacancy in their membership for a "legal beagle," I joined and almost at once was saddled with the job of secretary. This involved writing a weekly letter and having it printed and mailed to each member. There was also the Canadian Club with members from the town and surrounding county. I joined and within a year was appointed corresponding secretary. This involved contacting and lining up prospective speakers as suggested by the committee, and usually meeting them on arrival at the station or in their private railway cars, which many of them had. The Canadian Club and to a lesser extent the Rotary Club were the only means available in those days to meet or hear public figures. We had no television, no radio, no aircraft, and very few paved highways.

When I was in my senior year at high school, Robert Whyte, the principal, was president of the Canadian Club and used to invite about ten of us from the senior grade to come to the evening meetings after the meal to hear the words of famous men. At the end of the season it was suggested that one of us should thank the Club for giving us this privilege, and I was chosen to do the job. I gave the matter much thought and rehearsed my three-minute speech of appreciation to the members and also to the speaker for his very informative and interesting talk, which of course I had not yet heard! But I was sure it would be so, because the guest on this occasion was Newton Wesley Rowell, the leader of the Liberal Party in Ontario. On the appointed evening, I delivered my well-rehearsed words, and was thrilled to meet Mr. Rowell at the close of the meeting. He shook my hand and said, "Thank you, young man, for your very kind words. Some day I hope to see you in the political arena. You should do very well."

After joining the Rotary Club and the Canadian Club, I was accepted into the Masonic order. Here I did not become a secretary, but I soon was invited to "go through the chairs," which meant to start at the lowest office and proceed through the various positions or chairs in the lodge to finally become top man or "Master." This I did, though it took some seven years of faithful attendance and considerable study in Trent Lodge No. 38 on the Grand Register for Canada.

These and various other activities took considerable time. They kept me and my secretary busy and produced no direct remuneration. In an indirect way, perhaps, they made some contribution to the life of the community. At the same time they brought me in contact with people in all walks of life, which was good for business. It was not long before the law practice grew to the point where I felt it necessary to engage a young graduate named Alex Elliott, who opened a branch office in Frankford under the name of Graham & Elliott. Later I engaged another recent law graduate, Isabel Hinds, who was not only a qualified lawyer but also a first-class stenographer. Added to these, of course, were additional clerical staff and one or two articled students. Early in 1930, Alex Elliott left me to return to his native Peterborough and join in practice with G.N. Gordon, who was then Minister of Immigration in the King government. In Alex's place I was fortunate in getting a young graduate named Allan MacNab, who held the practice together for the duration of World War II. I was not keen about court work but would handle most of it when necessary. On rare occasions, on the criminal side, I would advise a client to authorize me to engage counsel, older and more experienced than I, in jury trials.

One of my cases demonstrates what I have always maintained, that a trial by jury is far more likely to result in a miscarriage of justice than a trial by judge alone. One fine summer day an American couple were driving a large Cadillac touring car through Trenton at the proper legal speed, or much less, when they were struck on the side by another car with such violence that his large car was knocked over on its side and the driver and his wife were thrown out. Fortunately neither was seriously hurt, except for one broken shoulder, because they landed on a soft grassy boulevard, but the car was severely damaged. The offending car was being driven by a little lady, the wife of a hotel keeper in Trenton, who, it was later learned, had had a tiff with her husband. In a fit of temper she had raced off in their car at such a speed that she could not make a proper turn at a bend in the road and so had slammed into the Cadillac. Her own car was damaged in front but she was not hurt. From the above facts, one can see that the American driver had a clear case for damages against the hotel keeper who owned the car his wife had been driving. Someone sent the American to me and we claimed an amount that included repairs to the Cadillac, out-of-pocket expenses for hospital and doctor's bills, and a reasonable sum for the pain and suffering of the driver's wife, who had a broken shoulder. The hotel keeper's insurance company contested the claim and in due course the case came to trial before a Supreme Court judge. Of course, the insurance company's lawyer, Thomas Phelan from Toronto, asked for a trial by jury. The jury was duly selected and most of the members were farmers, labourers, or local businessmen.

My client's name was Bozo Rankovic, who happened to be president of the Century Bank of New York City. (I assumed that it was a small neighbourhood bank.) No mention may be made before the jury that the defence is being conducted by an insurance company and that any damages assessed would be paid by that company and not by the defendant named in the action. Tommy Phelan, being one of the most astute and experienced jury-trial lawyers in Ontario, made the most of my poor client's nationality. Here was a foreigner who was driving a very expensive car, a bank president to boot, with a name like Bozo Rankovic, which Phelan repeated at every possible opportunity. A justice sitting alone would surely have ignored these unimportant facts and would have assessed damages "to fit the crime." Not so a jury of local farmers and small businessmen. We got minimal actual out-of-pocket expenses, a pittance for Mrs. Rankovic's pain and suffering, and nothing for delay and inconvenience on their holiday. My client was a good sport. He commended me on my part in the preparation and conduct of the trial and said he thought that our courts were conducted with much more dignity and decorum and courtesy than was the case in the United States. He was disappointed in the result, but held no animosity against either Tommy Phelan or the jury. However, I do not think he ever again visited Canada!

Over the years I became increasingly active in municipal politics. A Parent-Teacher's Association for the two public schools in town had recently been formed, and I was urged to stand for election to the Public School Board, which I did, and was duly elected and served for two terms. Near the end of my second term, I was asked by some of my business friends to run for membership on the Town Council. Trenton, being a "separated" town, i.e., not a part of a township, had a council or governing body of a mayor and six councillors. I made a successful run for the year 1927 and for succeeding years until 1933. I suppose I had achieved the distinction described by my former boss of being "a big toad in a little puddle." During most of those years in Council, I sat as chairman of the Court of Revision, which heard appeals by citizens with regard to the amount for which their property was assessed and hence the amount of taxes that would be levied against them. As a practising lawyer, this sometimes brought me into an embarrassing position vis-à-vis clients, even very valued ones. However, I tried to be fair in my decisions and explain the reasons for them. I can recall no instances of animosity arising from negative decisions.

In those days, perhaps more than now, political affiliations were important. My family had always been of the Conservative persuasion, and in the late 1920s I became president of the local Conservative Association. William Fraser, the mayor during my years on Council, was a very strong and prominent Liberal. He represented the adjacent riding of East Northumberland in Ottawa and eventually was appointed to the Senate. He owned the Trenton Cooperage Mill, which provided the many thousands of barrels required each year for the apple orchards. He was a good Grit and I was a well-known Tory. When we got to working together on town affairs, our respective political loyalties assumed little importance and we became fast and mutually trusting friends.

One day he phoned from his office at the mill and asked if I could come to see him about a town matter on which he wanted my opinion before he brought it up in Council. I found him at his desk with a file of legal documents. They referred to the transfer of Trenton's water rights in the Trent River to the Ontario Hydro Commission. I have forgotten the details of the deal that had been made before our time in Council, but we both agreed that the town had been grossly lenient in transferring water-power from that part of the river flowing through the corporate limits of the town. As a result, we discussed the matter in Council, and it was agreed that the mayor and I, as a solicitor, should pursue the matter in an effort to have an adjustment made that would be fair and of advantage to the town. This we did by meeting and negotiating with Charles Magrath, then chairman of the Ontario Hydro-Electric Power Commission. There were a number of meetings and though it took time from my practice, I enjoyed the discussions and arguments and particularly the delicious lunches that Dr. Magrath provided in his private dining room. Finally, by consent of Council, a settlement was reached whereby Trenton received an additional cash payment of approximately $50,000, no mean amount in the late 1920s and early 1930s.

A much larger and more important project of great benefit to Trenton had its beginning and culmination during my years in Council. It was largely due to the personal drive, energy, influence, and persistence of Bill Fraser supported by his colleagues in Council. The project I refer to was the location of the principal air training base for the newly formed Royal Canadian Air Force in the Township of Sidney on the eastern outskirts of Trenton. In the late 1920s, it was considered that hydroplanes would play a large part in the future of aviation. For that reason it was judged essential to have the air training centre border on a suitable body of water. The Bay of Quinte seemed to fit the bill because of its adequate size, depth, and configuration. There was also the fact that the essential land area to the north of and adjoining the bay was flat but well-drained and sloped gently to the shoreline. Other communities, particularly in the Lake Huron and Georgian Bay areas of Ontario, were bidding furiously for the air base. An MP as well as mayor of Trenton, Fraser undoubtedly used his very considerable influence to bear upon the Liberal government of the day to decide that the Trenton site was the most suitable. Though the site was outside the corporate limits of Trenton, the town was much involved in guaranteeing such services as the supply of adequate drinking water to the property, which was designed to include and does now include accommodation for a large number of Air Force personnel, both married and single.

The boundaries of the site for the air base being settled, the federal government purchased or offered to purchase from their owners the many properties that had to be assembled – a total of more than 1,200 acres. Almost all the owners, some of whom were clients of mine, accepted the government's offer and were duly and promptly paid. But eleven owners refused. Their properties were expropriated and their claims were referred to the Exchequer Court of Canada for adjudication. Nine of these eleven owners brought their cases to me, and the other two owners retained Dick Ponton, a Belleville lawyer whose father had befriended me when I had enlisted. These cases required a great deal of work, and I personally gave each of them very careful preparation. In deciding on the value claimed for each of the nine properties, many factors had to be considered and would have to be proven to the satisfaction of the Justice of the Exchequer Court, who would hear the claims. For example, how would you determine the value of an apple tree? What is a dairy herd worth? How do you assess the value of a cattle pen? They had to be valued by experts. Each case being different from the others, it required some months of research and consultation with owners and witnesses before we were ready for trial. Eventually, after many months, all was ready and Mr. Justice Audette of the Exchequer Court came from Ottawa to hold his sessions, which extended over a number of weeks, in the courthouse at Belleville.

All, or almost all, went well. His Lordship was patient and courteous and in due course rendered judgement, fixing in eight of my nine clients' claims substantial increases beyond the amounts offered by the government, plus interest on the full amount for the lengthy time the owners had been denied the use of their money and property plus costs against the Crown. Regrettably, one of my clients was not successful. He was a loyal supporter of the Liberal Party and he sincerely believed that the Tories had "done him dirt." Before the cases were heard, the Liberal government of Mackenzie King was defeated by the Conservatives under R.B. Bennett, and my client thought that politics had influenced the court's judgement. This was pure nonsense. Mr. Justice Audette was above reproach. Yet I must confess that I thought my ninth client did get a fair price for his property.

The stock market crash occurred in October, 1929, and very few people or businesses escaped its ill effects. Government assistance was minimal and confined principally to trying to ease unemployment by expenditures on government construction, which included work at the newly acquired Trenton Air Base. But government funds were limited. A policy instituted by the Bennett government provided work, food, and lodging for a limited number of single men and a cash payment of twenty cents a day at the air base. I was on the Council and we did our utmost with the funds available to keep people housed, clothed, and fed. Food vouchers were issued; churches, lodges, the Rotary Club, the Royal Canadian Legion, and other volunteer groups did their best to help support the less fortunate. In these circumstances I fell heir to the office of mayor. I was elected by acclamation for the year 1933. I think no one else would take the job! During that year a new industry was established in Trenton. The Downs-Coulter Company, a well-known, textile-weaving business in Bradford, England, decided to establish a factory in Canada. Largely through the generosity of ex-Mayor Fraser, who donated the site for the

89

new factory, they came to Trenton. The building of their factory and subsequent operations eased the unemployment somewhat. The Downs family were good citizens, and my wife and I got to know Mr. and Mrs. Downs, Sr., very well.

To provide the best food possible for the money available, the Council opened its own food store and stocked it with meat purchased by the carcass and cut into proper pieces by volunteer experts. Other staple foods, such as potatoes and root vegetables, were bought by the bushel, rolled oats by the barrel, and flour, salt, milk, and tea in quantity. Lady volunteers did the weighing and packaging and honoured the vouchers, all in all saving the taxpayers much money yet providing those in need with the kind of food they required. This was not a popular arrangement with many of the recipients. Quite understandably, they felt it was demeaning for them to get their relief in this way. They much preferred freedom of choice as to place and kind of food they should get with their vouchers. Furthermore, it was a policy of the Council that, if able to work, a recipient should do something for his "handout" and therefore he was obliged to do work on road repairs or other jobs that he often felt were beneath his dignity. One or more members of the Council knew most of the recipients of relief personally, and we knew that some of them did not like to do work of any kind. They were the most vociferous in their complaints!

I was glad to see the end of my term as mayor in that most difficult year of 1933, which marked the end of my service in elective office. I had, during nine years, devoted untold hours of time and professional skill to community affairs. During these years not one cent was spent by the School Board or the Council on legal advice. My professional services were gratis, and in those days no stipend or salary was paid to members of the Council or School Boards.

The most important event of my entire life occurred on June 1, 1926. My dear Jean and I were married. My bride came from somewhat the same type of background as I. Before Confederation her parents, William and Harriet Lowe, left Hamilton, Ontario, and settled on a virgin tract in the Township of Ryde in the District of Muskoka. It was there and in the town of Bracebridge, where her parents moved in later life, that Jean received her education, first in a country school and later in the town school at about the same time that I was going through the same experience in Trenton. On her graduation from high school, Jean enrolled as a nurse-in-training at the small but very select and efficient

Wellesley Hospital in Toronto. It was while Jean was in training that I met her on the blind skating date, and we both graduated in our respective professions in the spring of 1921. Jean did private-duty nursing in Toronto until 1925 when, with two others of her graduating class, she went to New York and continued private nursing until just before our marriage.

We were married in a small United Church at Clarkson, five miles from our present home, by her brother-in-law, the Rev. George Lawrence. After a ten-day motor trip (in a brand new Chevrolet car I had just bought), we settled in an attractive and comfortable home, which I had purchased two years earlier, within walking distance of my office in Trenton. We were very happy and over the years made many memorable trips together. In 1931, we attended the annual meeting of the Canadian Bar Association, which was held at the Manoir Richelieu below Quebec City on the shores of the St. Lawrence River. The motor trip took us through the Adirondack Mountains to Boston, then on a large seaworthy ferry to Yarmouth, Liverpool, and Halifax, where we saw fish piled near houses to look like cocks of hay with tails to the centre and heads to the outside, and more fish hanging by their tails on clotheslines. I was told they were cod being dried for winter use. We drove through the lovely Annapolis Valley to Digby, then by ferry to Saint John, New Brunswick, north to Rivière-du-Loup on the South Shore of the St. Lawrence River, by ferry across to Tadoussac at the mouth of the Saguenay River, and from there by a narrow, winding, and hilly road along the North Shore of the St. Lawrence to Murray Bay, where was situated the Manoir Richelieu. I faithfully attended the sessions while Jean enjoyed entertainment arranged by the ladies' committee. It is of interest to note that the president of the Canadian Bar Association that year was a young Quebec lawyer named Louis St. Laurent, later to be Prime Minister of Canada. Forty years later I had occasion to sit beside him at a luncheon in Ottawa and we enjoyed a half hour of reminiscing about those bygone days.

On our return to Trenton from Murray Bay, Jean and I went through Quebec City and stopped for an hour to visit Wolfe's Cove and sightsee aboard the recently built Canadian Pacific *Empress of Britain*. What a beautiful ship she was. We agreed that some day we would like to cross the Atlantic in her. That "some day" turned out to be the very next summer. It so happened that I was acting during 1931-32 as the Canadian agent for a firm of solicitors in London, England, who were acting for the executors of an estate, the assets of which were being distributed to heirs living near Trenton. This involved a considerable amount of legal work. I suggested to Jean that this might be a good time to make our hoped-for voyage in the *Empress of Britain* and Jean's first visit abroad. We read the recently published books by H.V. Morton, *In Search of England* and *In Search of Scotland*, and decided to follow his trail, so to speak.

Early in the summer of 1932, when most people were struggling to make a living, I was fortunate to have available cash, not as the result of a gift or a legacy but because of diligence and long hours of work at my practice, to finance the trip we had planned. Early in June we boarded the *Empress of Britain* (tourist class) at Wolfe's Cove in Quebec City and sailed for Cherbourg. I wanted Jean to see some of the sights that I had seen fifteen years earlier in and around Paris. The *Empress* was a fast ship, and after leaving the Gulf of St. Lawrence we had only four days on the open sea before dropping anchor off Cherbourg. Here we went ashore in a lighter and then by boat-train to Paris. There we stayed at a charming small hotel, The Windsor, near the Place de l'Etoile. There were few travellers, and the cost of accommodation and meals was very reasonable. During the next week we did indeed see all the sights. Then we flew in a great four-engined British Airways Hannibal aircraft from Orly to Croydon, which was then the only London airport. There were only about a dozen passengers. At the front of the cabin, above the door to the cockpit, two large dials like clock-faces showed the air speed, which was about 130 mph, and the height, about 3,000 feet. The flight was not of long duration, but as soon as we were airborne a steward in crisp white jacket came in and put up a table between us (our seats faced each other), covered it with a snow-white tea cloth, and brought a silvered teapot of tea, cream, sugar, and biscuits. The other passengers were treated likewise. It was a very pleasant and commodious way to travel. We came down at Croydon on a grass runway.

In London we stayed at the Victoria Hotel on Northumberland Avenue, a stone's throw from Trafalgar Square. I made my call on the solicitors and closed out the business I had been doing for them. After showing Jean the sights of London and its environs, we picked up our rented touring car, an Austin 12, the top of which folded back, and set forth on the road to Brighton and the tour that would take us to Land's End, at the southern tip of England, and thence north to Inverness in the northern region of Scotland. Our travels took us through the beauties of the English and Scottish countryside, and, apart from Edinburgh, I recall that we spent a night at only one other city, namely Nottingham. Traffic on the roads was light. We travelled on roads, narrow and winding, through forest and heath, through the Lake Country of England and around the lochs of Scotland. In most country inns, where we spent the nights, the tab was one guinea for bed and a full breakfast and shoes shined if you left them outside your bedroom door!

Back in London a few days before we were due to sail for Canada on the *Duchess of Bedford* from Liverpool, we again stayed at the Victoria Hotel. I said to my wife, "You know, I must go over to Canada House and pay my respects to the High Commissioner, or Bill Ireland will never forgive me." Why did I say this? Bill was the mayor of Trenton when I was discharged from the army, and he later became our provincial member of Parliament and was Chief Conservative Whip when Howard G. Ferguson was Premier of Ontario. When Bill learned that I was going to England, he said, "Howard, be sure and call on Mr. Ferguson, the High Commissioner. I will write to him and tell him you'll be in to see him."

I was not too keen about imposing my presence on the High Commissioner, but having the morning free and Canada House being only a couple of hundred yards from our hotel, and not wanting to disappoint my friend Bill, I went across Trafalgar Square and presented myself to Mr. Ferguson's secretary. She welcomed me. "Oh, yes, Mr. Graham, we had a letter from Mr. Ireland about you. Please sit down and I'll tell the High Commissioner you are here." She did, and in a moment Mr. Ferguson, whom I had never met before, welcomed me warmly in his large office overlooking the Square. (I was to be a frequent visitor to this office a quarter-century later, when I was Senior Canadian Army Officer in London and adviser to the then High Commissioner, Norman Robertson.) As I was about to leave, Mr. Ferguson said, "By the way, would you like to attend the garden party at the Palace this afternoon?"

"Indeed we would," I replied.

"Well," he said, "sometimes invitations are returned at the last minute, and if this happens, I'll have my secretary phone you."

We stayed close to the phone and within the hour the secretary phoned to say that if we did not mind being Colonel and Mrs. Armour from Toronto, we could have their invitations, as they had been returned because the Armours were unable to attend. Off we went post-haste in a taxi to Harrod's and bought a beautiful big floppy frilled chapeau. In mid-afternoon we were in one of the long line of taxi cabs crawling up

93

the Mall to the Palace entrance. Our invitations were checked and we were ushered through the halls to the large, walled, twenty-five-acre garden at the back.

There was a tendency for the guests, eventually many hundreds, to keep close to the terrace onto which their Majesties, George V and Queen Mary, and other members of the royal family would emerge through French doors. I recall the scene vividly. On the terrace, awaiting to attend and escort their Sovereign, were three tall equerries from the Indian army in full-dress uniforms, complete with turbans, knee-length and brightly coloured (but all different) tunics, with decorations aplenty. In due course, the French doors were opened and the members of the royal party appeared, men in formal morning attire, ladies in long gowns, and made their way slowly, the King one way, the Queen another, the Prince of Wales another, stopping to say a word or two to some of the guests. We were not so favoured until they reached the royal enclosure where they formally received certain dignitaries and then, under an open-sided marquee, had tea with them. The rest of us enjoyed a delicious tea, served under a great long, open-front striped canvas shelter, with strawberries, cream, and cakes. It was a beautifully warm and sunny July day, the gardens glorious with flowers at their best. Although it was a thrilling experience, and though we have been to a number of royal garden parties since then, that one in 1932, being the first, and so unexpected, remains fresh in our memories. So ended our travels, which emanated from my attendance at the Bar Association meeting the year before!

Being the president of the Trenton Conservative Association, I was a delegate to the Conservative convention in 1937 in Ottawa, which would choose a new party leader to succeed the Rt. Hon. R.B. Bennett, Prime Minister during the difficult Depression years from 1930 to 1935, who was now resigning. His government and the Conservative Party had been badly beaten at the polls in 1935 by Mackenzie King and the Liberals. A good number of the delegates from Ontario were well known to me, and we gathered in the Château Laurier and exchanged views on what the future policies of the Conservative Party should be and who the new leader should be. On the final ballot the choice was between Dr. R.J. Manion of Fort William, Ontario, and Murdoch MacPherson, a highly regarded lawyer from Regina, Saskatchewan. I knew neither of these gentlemen personally, but I was much impressed by MacPherson's personality and platform manner and, above all, by the reasonable and sensible policies he advocated in his nomination address.

My friends did not deny or oppose my views on those points, but they contended that a leader from the Prairies would not have a chance at the polls in an election to secure seats from eastern Canada and particularly in Quebec, where in 1935 the Tories had been practically wiped out. They contended that Dr. Manion, having a French name, speaking the French tongue, and being a Roman Catholic was the man we should have if we were to win a badly needed bloc of seats in Quebec. My answer to these arguments was clear and unwavering. We should choose the leader who had the best qualities of leadership, the best presence, the best personality, the best ideas, and the ability to promote them. These assets were of infinitely more importance and real value in the battle for seats than the superficial values that my friends were attributing to Dr. Manion. I recall that the discussion went on into the early hours of the morning of the day of the final ballot. The views of my friends were shared by the majority of delegates. Dr. Manion became leader.

I was not happy with the choice of the new leader, but that was now a *fait accompli* so I would say no more about it. But I was also disappointed with the arid results of the convention, and when in the closing minutes the chairman asked if any person wished to say something before adjournment, I rose to my feet and said, "Yes, I have a few words to say." I thought he would say, "Go ahead, Mr. Graham," as we knew each other, but instead he invited me to the platform, introducing me thusly, "Ladies and gentlemen, Mr. Howard Graham, QC, former mayor of Trenton, Ontario."

My remarks were few but pungent and perhaps impertinent in the view of some. I said, "I came to the convention not only to aid in the selection of a new leader for this great old party, which had founded and fostered Confederation, but also in the hope of learning new ideas, of positive and forward-looking policies, of proposals for amendments to the Canadian Constitution that would help to heal the wounds that were still smarting among French Canadians. But I am leaving with profound feelings of disappointment. It is as if I came expecting a feast and instead was given a few crumbs." That was a dreadfully sour note on which to close the convention. The chairman simply said, "Thank you, Mr. Graham," and there was a smattering of applause. Perhaps I should have kept my mouth shut, but it may have given the Executive Committee something to think about.

My next foray into the political arena came on April 5, 1939, when the South Hastings Conservatives met in Belleville to select a candidate to contest the federal seat in the next election. I had been president of the Trenton Conservative Association for four years and was quite well known throughout the riding. My friends in Trenton and vicinity pressed me to make a bid for the nomination. Four other men had announced their intention of running: Jamieson Bone, mayor of Belleville; Arthur Burke; George Stokes, former sheriff of Hastings County; and Dr. W.C. Morgan. All were of the city of Belleville. I had no desire to be a federal member of Parliament. I owned my own home. I had a devoted wife, and now a son of six years. I had a successful and profitable law practice. If I were nominated, it was generally conceded that I was almost sure to gain the seat at the next election, expected to be held the following year. If elected, I would receive the stipend for a federal member of $4,000 *per annum*, which would mean a substantial financial loss. My practice would suffer. For long periods I would be away from my home and family, and I did not care a tinker's dam for the so-called honour of being an MP. However, my friends persisted and, although Jean did not demur, I felt that she would prefer that I reject the bid. I finally said, "Okay, I'll run, but I will not personally canvass for support, nor will I contribute any money to the contest." Nor did I. As I look back I am somewhat ashamed of my lack of graciousness. Most certainly, if I had been nominated and elected in the forthcoming election (as I think I would have been), the pattern of my life would have been different, whether for better or for worse is hard to say.

The *Ontario Intelligencer*, the local Belleville daily, in reporting the Conservative nomination meeting on April 6, 1939, described it as one of the most vigorous and hard-won elections ever held. Indeed it was. Each of the five delegates who allowed his name to be placed on the ballot was allowed five minutes to address the meeting. I spoke off the cuff, and the paper reported me as saying, in part: "I have been at times a cantankerous and critical Conservative. I'm not offering myself for the job for the $4,000 a year it pays or for the headlines in a newspaper. I am trying to give the same sort of support to the party and the country that I was able to give twenty years ago when the country needed me." The paper also reported that I said that I had not canvassed the delegates because I felt they should come to the convention with an open mind and unfettered hands. After these and other feeble and futile remarks, it is surprising that I survived the first and second ballots. On the third ballot, with only George Stokes and myself left in the race, George won by a majority of

(ABOVE RIGHT) Jessie MacPherson
Graham, the author's mother,
taken about 1890. (ABOVE) Wesley
Gilmour Graham, the author's
father, about twenty-five years of
age. The photograph was taken
prior to the move to Buffalo in
1890. (RIGHT) Grandmother
Sarah MacPherson with the
author's sister, Hobie, at age
three.

(LEFT) Three generations of Graham women, in a Trenton photo from 1905: the author's grandmother, Sarah MacPherson; his mother, Jessie Graham; and his sister, Edith. (ABOVE) Howard Graham, age eleven, with his sister Edith – "all tidied up and ready for Church"! (BELOW) The house the Grahams built, begun in 1909, finished (as above) by 1912; it still stands in the Dutch Settlement, Trenton.

(RIGHT) Private Graham, H.G., Serial Number 636774, the newest recruit in the Canadian Army, March, 1916. (BELOW) With the army at headquarters, Bonn on the Rhine, Christmas, 1918. The author is shown standing, sixth from left.

(TOP) Wedding party. Jessie Lowe and Howard Graham were married at Clarkson, Ont., June 1, 1926. (ABOVE) King George VI is being escorted through the lines of the Hastings & Prince Edward Regiment in Aldershot, England, January, 1940. Accompanying His Majesty are Major-Gen. Andrew McNaughton and the author.

General Sir Bernard Montgomery pins the ribbon of the D.S.O. on the left
breast of Brig. Graham to honour action undertaken "with complete
disregard for personal safety" in Sicily on July 18, 1943.

Shown are the Brigadiers of the 1st Canadian Infantry Division, a mile or two south of Massena, Sicily, August 31, 1943. From this point could be seen, across the Strait of Massena, the future landing places in Italy. Left to right are: Howard Penhale, 3rd Brigade; Chris Vokes, 2nd Brigade; Bruce Matthews, Royal Artillery; and Howard Graham, 1st Brigade.

Senior Officers of the 1st Canadian Infantry Brigade, near Mount Etna, Sicily, August 22, 1943, all wearing short pants! Left to right: Lieut.-Col. Robert Irvine, c.o., Saskatoon Light Infantry, later killed in action; Lieut.-Col. Ian Johnston, c.o., 48th Highlanders; Brig. Howard Graham; Lieut.-Col. Bert Kennedy, c.o., Hastings & Prince Edward Regiment; and Lieut.-Col. Dan Spry, c.o., Royal Canadian Regiment.

(ABOVE) The new Chief of the General Staff, Howard Douglas Graham, C.B.E., D.S.O. This photograph is reproduced from *Saturday Night*. (OPPOSITE, TOP) "General Laughter" was the caption this photograph bore when it first appeared in the *Illustrated London News*. It shows the author with Field Marshal Sir Gerald Templer at a conference of the Chiefs of the Imperial General Staff, Camberley, England, August 9, 1956. (OPPOSITE, BELOW) With Governor General Vincent Massey and Prime Minister Louis St. Laurent at the Opening of Parliament, January 8, 1957.

Happy Soldiers: Lieut.-Gen. Howard Graham, C.G.S., and Major-Gen. John Rockingham, G.O.C., 1st Canadian Infantry Division, Gagetown, N.B., July 1, 1957.

(TOP) The day Field Marshal Montgomery came to take tea at our home in Ottawa, May, 1956. (ABOVE) The Army Council, May, 1958. Left to right: Major-Gen. G. Walsh, Q.M.G.; Major-Gen. M.L. Brennan, A.G.; Lieut.-Gen. Howard D. Graham, C.S.G.; Major-Gen. H.F.G. Letson, Militia Adviser, A.D.M.; Major-Gen. George Kitching, V.C.G.S.

(LEFT) High Society. Government House reception, Ottawa, 1959. (BELOW) Tall in the Saddle: Royal Visit, Ashcroft, B.C., July 14, 1959.

Members of the Canadian group who travelled with Her Majesty the Queen on the Royal Tour of 1959.

Her Majesty Queen Elizabeth and His Royal Highness the Duke of Edinburgh, accompanied by her three Canadian Equerries, her Canadian Secretary, her Private Secretary, and her two Ladies in Waiting, on the dais in front of the Peace Tower, just prior to the reading of the Centennial Address, July 1, 1967.

(ABOVE) The Boy Scouts of
Canada and Les Scouts Ca-
tholiques du Canada became
affiliated on February 22,
1967. Seated left to right are:
the author, who brought
about the association; Governor General Georges Vanier,
by virtue of his office Chief
Scout; and M. Jean-Marie
Poitras, President of Les
Scouts. The signing took
place at Government House,
Ottawa, with representative
Scouts, both English and
French, looking on. (RIGHT)
The End of the Road. Jean
and Howard in the garden of
their Oakville home, overlooking Lake Ontario, July, 1982.

The General 1967.

five votes on 277 votes cast. I was quite happy with the result, and a few days later Jean and I, leaving our little son with his aunt in Toronto, drove to Atlantic City for the Easter holiday.

It is of interest to note that this nomination meeting made the headlines in the prestigious Toronto weekly *Saturday Night*. The editors wrote in the issue of April 15 that the nomination meeting "was a red-letter day, or perhaps a black-letter day in the history of the hustings of Hastings, Ontario. . . ." The editors continued:

The district has been fairly sizzling of late with Mr. George McCullagh's brave words on leadership, Government excessive taxation, unemployment, and welsh rarebit [a reference to a well-known Hastings product – cheese!]. Here was an opportunity to show the public at large that South Hastings, at any rate, was determined to have a Government by the people and for the people. One candidate was a lawyer with an excellent war and civil record. One was a doctor with a similar record, one was Mayor of Belleville, who had been quoted from coast to coast for his daring utterances on major problems along leadership league lines.

All these went into the discard and the 286 delegation loudly applauded and chose as their standard bearer Mr. George Stokes, a retired farmer and life-long Conservative. There appears to be nothing wrong with Mr. Stokes and we are all for him, but we were very much interested in his selection because he apparently has what it takes, and what is it? And how is the leadership idea taking root? Here is the answer. All candidates for political honours please take note.

Mr. Stokes' father was a defeated Conservative candidate many years ago. Mr. Stokes himself was almost chosen Provincial Conservative candidate almost 25 years ago. For 16 years he served faithfully as Township Clerk of Rawdon. About seven years ago he was appointed Sheriff of Hastings County, by the Provincial Government, but when the Liberal Party came into power about two years later, he was replaced by a good Liberal. Mr. Stokes has never frittered his time away with property, manufacturing, or Chamber of Commerce problems of the day, or with anything other than the Conservative party. Do the Conservatives of Hastings really expect him, if elected, to forget party and work for the common good?

Somehow the above seems to endorse and indicate what I tried to put across at the Conservative convention in Ottawa in 1937.

One evening in early June, 1939, the phone rang and it was Kenneth Couch, the mayor of Trenton. "Howard," he said, "I have a man here I think you should see. He has a confidential matter to discuss."

I said, "Okay, Ken, bring him up to the house. We can talk about it here." In about ten minutes Ken arrived and introduced John Suchanek, a Czechoslovakian who represented the Bata Shoe Company.

I had never heard of Bata Shoe, but I soon learned from my visitor that the patents for shoemaking machinery were then owned by a very small number of people. Most shoe manufacturers throughout the world were obliged to buy or rent their equipment from one of these people, one of whom was Tomas Bata, a native of Czechoslovakia. Suchanek emphasized in the greatest confidence that the Bata Company was about to remove its equipment from the homeland and had decided to establish a North American base in Canada. He went on to say that the company had studied many locations and that he and others thought that the area between Trenton and the village of Frankford, extending inland from the west bank of the Trent River and the Trenton-Frankford Road, which ran along the river's edge, would be most suitable. I knew the area well. In fact, our old family homestead, settled by grandfather Andrew Graham, was not far away.

Suchanek then unrolled a map, which I at once recognized as a One Inch Ordinance Military Chart, which is on a scale of one inch to one mile and shows locations of houses, barns, schools, churchs, roads, concession lines, farmsteads, wooded areas, orchards, railways, etc. On this map he had outlined the area that the Bata Company was interested in, and it contained more than 1,500 acres owned by many farmers, having acreages varying from fifteen to several hundred. I have forgotten the exact number of landowning farmers, but I think I could have called 90 per cent of the owners by their first names. The task was to assemble this block of land within a very limited time, about three months, without the advantage of the right to expropriate, which the federal government used when it acquired a similar-sized area for the air base east of Trenton in 1929. The task was mine. The question was how to accomplish it and maintain secrecy as to the buyer and how to avoid holdouts, i.e., owners who would want an unfair and exorbitant price for their properties.

Fortunately, the Bata Company's lawyers in Toronto, who would be arranging Canadian incorporation and other matters, were well known to me. The member of the legal firm who would be in charge of these matters was Wilfred Perry, a veteran and personal friend. Wilf agreed with the plan I evolved, which was to acquire options to buy each and every property at a fixed price, the options to be taken in the names of friends of mine to be exercisable on or before August 31, 1939, and after that to be null and void. I would do the negotiating with the owners. On this basis I enlisted the help of eight citizens, mostly businessmen who had served with me on Council, including William Bain, the chief constable of the town, a neighbour and close friend. I was fortunate in having with me in the office two junior barristers, Isabel Hinds and Allan MacNab, and a student who eventually bought and carried on the practice himself, Robert Smithrim. At the time of these proceedings, I had not met any member of the Bata family, but soon afterwards I had the pleasure of knowing intimately young Thomas Bata, who now lives in Toronto and is the head of the world-wide Bata Shoe manufacturing and marketing empire. All went according to plan. The options were duly obtained on all the properties. The last one to be exercised was completed on the morning of August 31. The Bata Company has prospered, and Trenton and the surrounding hamlets have benefited greatly from its operations. The area that was purchased in the summer of 1939 soon became a thriving, self-contained community and now carries the name of Batawa and has its own postal code.

Near the end of that beautiful August day, as I was clearing my desk to go to the cottage with the feeling that a difficult task had been accomplished, the phone rang. My secretary said, "Colonel Young is on the line."

I picked up the receiver and said, "Hello, Sherman, what's new?"

His reply was short and shattering. "We are mobilized as of one minute past midnight tonight. Get into your uniform."

Colonel Young's telephone call took me by surprise. My involvement with the Militia went back to World War I. Early in 1923, as I was leaving the Hotel Gilbert in Trenton after a Rotary luncheon with Arthur Baywater, a fellow Rotarian, he said: "Howard, why don't you come into the new regiment that I am just getting organized? I would be glad to have you and there would be no problem in getting you a commission."

"That sounds interesting, Arthur," I replied. "I'll think it over and give you a call."

A.E. Baywater was a veteran of World War I and had been severely wounded. After the war the Non-Permanent Active Militia, as it was called, was reorganized. Before the war almost every county in Ontario had a County Regiment. The number of regiments was in 1920 reduced by a program of disbanding or amalgamating units. So it was that the 16th Prince Edward (County) Regiment and the 49th Hastings Rifles were amalgamated to form the Hastings & Prince Edward Regiment, and Major Baywater, as he then was, was promoted to the rank of lieutenant-colonel to command the new unit.

It did not take me long to think it over. If I was going to be a "joiner," why not join the local regiment and be, for the first time, a commissioned officer and get to know and be associated with others of my own vintage from all sections of the two counties? So I called the Colonel, told him of my decision, and on May 13, 1923, was gazetted a provisional lieutenant on the strength of the Hastings & Prince Edward Regiment. In August of the following year, in Picton, I took a short course of instruction and became a full lieutenant. Five years later, after attending a camp school of instruction at Barriefield near Kingston, I passed the required examination and became a captain in January, 1930.

In order to qualify as a "field officer," i.e., a major and above, it was necessary in those days to pass a Militia Staff Course. Thus I attended a course of evening lectures throughout the winter of 1931-32 given by a permanent force officer who came from Kingston to Belleville one night each week. By 1931, the road from Trenton to Belleville was paved and kept reasonably clear of snow, so that each Thursday, after an early evening meal, I drove along the twelve miles to the Belleville Armoury to join some twenty or thirty Militia officers from my own and other regiments and heard lectures on such military subjects as organization, administration, military law, tactics, map reading, and training. After a day's work at the office or in court, it was something of an ordeal. On some winter nights, when it was snowing and blowing and drifting, it was a most tiresome journey to Belleville and home again near midnight. However, I persisted, and in early spring spent two days in Kingston writing examinations on what was termed the theoretical part of the staff course. These tests were not difficult and I had no trouble passing.

The next stage of the staff course, called the practical phase, involved attending a two-week course at Petawawa Camp during the following summer. Since I was travelling in Europe in 1932, I deferred taking the practical phase until 1933. Then I attended and, as my certificate states, "having passed the required tests, is qualified for appointment to the staff of the N.P.A.M. of Canada in all grades up to and including Brigade

Major." Hence, on August 23, 1934, I was gazetted major after ten years
Militia service and after passing the Militia Staff Course.

I learned a little lesson at Petawawa from Major Bradbrooke. One of
the tests on the practical side of the staff course was to prepare a plan to
carry out successfully a military operation based on certain facts given to
us by our instructor. For example, it might be for a withdrawal of a
brigade while in contact with the enemy, or it might be to prepare orders
for an attack over certain terrain against an enemy position. In this
instance, we were a syndicate, a group of eight officers, and I was the
syndicate leader. The problem was to prepare an operation order for a
battalion attack on enemy positions. We gave the problem much
thought, and I wrote the order in great detail with their help. In due
course we presented it to "Brad." He read it over carefully, thought for a
few moments, and then gave one of the best critiques I had ever heard.
"Um," he said. "Very good gentlemen, it might work. But, you know, it
reminds me of a whisky and soda with a hell of a lot of soda and very
little whisky!" And I must confess he was quite right. He taught me that
in preparing and giving orders, it was quite unnecessary to embellish
them with flowery phrases. There should be no unnecessary verbiage to
clutter up the essence of the subject. In later years I heard the story that
General George Marshall, when he was American Chief of Staff, would
return to a subordinate officer any appreciation or opinion on a particular subject or situation if it covered more than one foolscap sheet.

Before I received the certificate referred to above, I had to pass a test on
equitation, or horsemanship. As a boy I had ridden any number of horses
at the walk, trot, gallop, and even jumped low fences, but always bare-
backed. Now, with two fellow officers, I had to go to Kingston, where the
Military Riding School was located, and do a few turns aboard a horse
with saddle. Very slippery it was! The examining officer was Captain
(later Major-General) Churchill Mann, an expert in riding who was for
several years a member of Canada's international jumping team.
"Church" had us go through the various gaits with and without stirrups,
but put us over no jumps, thank goodness! At the finish he said, "Well
done, gentlemen. I am glad to say all of you have passed and shortly you
will receive a certificate saying you are proficient in equitation. But one
word of advice, never show it to a horse!" Such was the state of Canadian
Militia thinking as late as 1933. Instead of riding horses, we should have
been taking tests in driving and servicing motor vehicles.

In 1935, it was announced that the memorial atop Vimy Ridge in France, which had been under construction for several years, was nearing completion and would be unveiled in the summer of 1936, nineteen years after the great Canadian assault and capture of the important strategic feature on the Western Front. It was the hope of the government and the officers of the Canadian Legion that a large contingent, perhaps as many as five thousand, would attend this important event. It would be called "The Vimy Pilgrimage," but there would be no government assistance in the cost of travel. However, during the three or four days that veterans would be in France, the French and Canadian governments would provide accommodation and meals. I was then a member (and am now a life member) of the Trenton branch of the Royal Canadian Legion (No. 110) and served as a member of the executive committee. Someone conceived the idea of holding a raffle or lottery to raise funds to pay the expenses of a delegation from our branch to attend the unveiling ceremonies. The lottery prize was to be a motor car, and tickets were to cost one dollar each. We had what I would call a strong executive committee, having as members two bank managers, the postmaster, a doctor, a lawyer, and two or three businessmen. With this group in charge, plans were quickly made. The cause was popular. Even though the Depression was far from over, people were prepared to help a number of Trenton veterans attend the Vimy unveiling. The response was terrific, so much so that instead of one we put up three cars as prizes. The dollars flowed from neighbouring towns, villages, and the city of Belleville, so that in the late spring of 1936, when we closed the operation and had the final draw for the third car, we paid our expenses – we got the cars wholesale – and there were funds enough to pay the travelling expenses of all Legion members who were able to make the trip. I was one.

On July 16, 1936, a convoy of five ships sailed from Montreal, crowded to the limit with more than six thousand pilgrims. After a pleasant and uneventful voyage on the s.s. *Montcalm*, we sailed up the Scheldt Estuary early on July 25 and berthed at Antwerp. From here we were taken by bus to various French and Belgian towns for billeting. Our group was put in a boys' boarding school in Armentières, which was comfortable and quite adequate. Very early on the morning of July 26, after a good breakfast and a box lunch, we left our school billet by bus to arrive mid-morning at the memorial and be assigned our place on the grassy slope that falls away eastward from the base of the monument. I will not try to describe in detail the magnificent memorial, designed by Walter Allward and built by Canadian craftsmen, which tops the ridge and commands a view from many miles to the eastward across the Douai Plain. We had a long wait, but the day was warm and sunny. We had many friends with whom to reminisce about the events of twenty years ago, and we had our box lunches!

Punctually at 2:15 p.m., King Edward VIII arrived. We could hear the trumpets sound the royal salute and the shouted orders to the Honour Guards of veterans, and soon thereafter, His Majesty with the large assembly of dignitaries – Canadian, French, and British – appeared upon the rampart. He waved his hat to us assembled below. Then followed the addresses and the formal unveiling of the cowled figure of a woman, beautifully carved in stone, representing Canada mourning for her dead. After the unveiling, His Majesty descended the memorial steps and walked through our ranks, stopping occasionally to say a few words to a man or woman who would be wearing a decoration, usually for valour. We were among the few Canadians who ever saw Edward as a sovereign. He abdicated the throne after a reign of only 325 days later that year. The King returned with his entourage to the west side of the memorial and departed. We did likewise, but many spent some time visiting and walking through the old trench systems.

The next day we sailed on the s.s. *Montcalm* from Amsterdam to the London docks at Tilbury, and thence to a small hotel, Dean's, in Oxford Street. The next few days were filled with events of a lifetime. A great reception was held in the ancient Westminster Hall, with speeches of welcome by Prime Minister Stanley Baldwin and others, followed by a solemn and moving service at the Cenotaph in Whitehall. On July 29, a garden party at Buckingham Palace was to be hosted by the Duke and Duchess of Gloucester because the King was expected to be in the south of France. But he postponed his departure and unexpectedly turned up at the garden party and, as at Vimy, mingled with and chatted with many of those who attended. Thereafter we returned to Canada.

In 1935 the Department of Defence had instituted an Advanced Militia Staff Course for a limited number of Militia officers, a dozen or so for all of Canada. This course was intended to equip those who passed it for senior staff or command positions in the event of war, just as the Militia Staff Course qualified graduates for appointments only up to brigade major. In the fall of 1936, a second advanced course was authorized and units of the Militia were invited to submit names of proposed candidates. My regimental commander urged me to allow my name to be submitted.

He argued that it would be advantageous not only to me personally but also to my regiment. It was a distinction to have one's name on the Militia List of Officers followed by the letters m.s.c. to indicate a pass in the ordinary staff course. But it was of much greater importance to have the M.S.C. in capital letters to show a pass in the advanced course.

I was loath to spend more time than I was already on military affairs, as it was at the expense of my profession, my livelihood. But I finally agreed to have him submit my name. I was accepted, and during the winter of 1936-37 I would receive every two or three weeks a large bundle of magazines and papers that were to be perused. On the basis of information contained therein, candidates were to submit articles in which they would comment, agree, or disagree with what we had read. The material to be read covered international political situations, natural resources of various countries, forms of governments, and, of course, the organization and command of large military formations. I did the reading in the evening and found it all very interesting. I sent in my articles as prescribed to the director of the course, Colonel Ken Stuart, who was later to become Chief of the General Staff.

Being a lawyer with some experience in clear expression of arguments, and in putting them on paper, I rather enjoyed reading and certainly learned a great deal about international affairs. I did not mind disagreeing with some of the views expressed in the material that came to me. For example, I recall learning about a Canadian proposal to recruit, arm, train, and equip a number of divisions and put them in the field as had been done in World War I. My response to this was that from my impression of the public's attitude toward defence, such a program would not be feasible unless conscription was introduced as soon as war was declared, unless industry was mobilized at the same time, and unless money was made available to equip and train the Militia with the modern weapons, equipment, and vehicles that would be required in mobile warfare as opposed to the fixed lines of trench warfare we had seen in World War I. In 1936-37 we had a permanent, full-time army of only some 4,000 to 4,500, all ranks.

The outcome of the winter correspondence course was that twelve officers from across Canada were chosen to take the practical portion of the Advanced Staff Course to be held at Royal Military College in Kingston in the summer of 1937. I was one of the chosen few who came from across Canada. Of the twelve candidates, five were lawyers, and of these one from Halifax was a Rhodes Scholar and another became a

Justice of the Supreme Court of Alberta. The course lasted a month. There was a directing staff of three officers. One, Colonel Stuart, became Chief of the General Staff; another, Major Burness, was lost at sea in 1940 when the s.s. *Nerissa* was sunk by u-boats; and the third, an English officer, Colonel Bucknell, then in Canada on exchange, became a British Corps Commander under Montgomery in World War II. I mention these names to indicate that I was in good company. We had first-class instruction and, as a result, much stimulative discussion.

So much about the courses of instruction in qualifying for senior army rank. But what of my regiment during the years from 1922 to 1939 and my part in its development? Colonel Baywater got us off to a good start by recruiting a full slate of officers from all parts of the counties of Hastings and Prince Edward. Some were farmers, some were young businessmen, and others came from the professions. Almost without exception in the early 1920s, all officers had served in France in the 1914-18 conflict. As the years passed, many of these men became too old for service and were replaced by younger men, but still, when mobilization came on September 1, 1939, the commanding officer and three or four others, including myself, were "veterans." Money was scarce and, as is always the case, as the period of peace after 1918 grew longer, the populace became less interested in matters military. Governments, yielding to the feelings of the voters and being exceedingly short of funds during the Depression years, cut the defence budget to the bone. Summer camps for the Militia at Barriefield and other central areas across Canada, where units from the various military districts had assembled for combined training, were discontinued for several years. Each unit, working on a very restricted budget, had to arrange for local camps. For example, one year the Hastings & Prince Edward Regiment had a camp on the fairground at Picton, in Prince Edward County, and another year at Madoc, in Hastings County, and others in Trenton and Cobourg. These were the only times that the members of a rural Militia unit like ours saw their fellows. In those days the officers brought their own uniforms and turned their pay over to the regimental fund. I can never recall drawing a cent of pay for time spent in regimental training.

We tried to keep the unit active and before the public in many ways. For example, in the 1930s we had several officers who were past masters of their respective lodges through the regimental area. When one of our officers was due for installation as master of his Masonic lodge, he and those of us who were past masters dressed in mess kit and we conducted the installation ceremony. On another occasion, an officer and his wife were having a baby christened in the Anglican church at Picton. We dressed in blue uniforms, complete with swords (all of which we bought out of our pockets), and, accompanied by our wives, attended the ceremony and afterward had a grand tea party in the Picton Armoury. I mention these affairs as examples to show how this regiment, and I am sure many others across Canada, kept the spirit and morale high in those difficult years, so that when the time came (as many of us were sure it would), this regiment would be able to take its appointed place in the mobilization plans. In a goodly number of years, the Hasty P's, as they came to be known, won the Infantry Association Cup for general efficiency in our Militia district.

What part did I play in this program? After qualifying as a lieutenant, I was posted to "A" Company of the regiment, centred in Trenton. Within two or three years, because of retirement or removal from the area of officers senior to me, I found myself commanding the company and continued to do so as I progressed in rank and qualifications. It was not until the late 1930s that I left "A" Company to become senior major and second in command of the regiment. I do not want to give the impression that commanding a company was a very onerous task. The company at any one time would not exceed a strength of forty. Trenton had no armoury, as such, but the Department of Defence leased a large barn, or shed, which had been used as a factory. Iron baffles were installed in one end, so that during the winter we could do "miniature rifle practice" using .22 ammunition fired from rifles that had been fitted with barrels with a .22 bore in place of the regular .303 bore. One corner of the shed was partitioned off as an office and a store room for uniforms and weapons. The bolts for rifles I kept in my office safe. I usually supplied the kindling wood for the stove, which burned soft coal, and often on a cold winter night in those distressing days of the Depression, Jean would make a basket of sandwiches, and with these and a handful of tea leaves we helped to keep up the morale. In the summer I might take as many as thirty-five men to camp, and since many had only well-worn canvas sneakers, I arranged with a local shoe merchant to provide a pair of shoes where needed on the receipt of a chit from me, on credit until after the camp and the men had received their pay. In many cases, of course, I ended up paying for the boots, but they cost only $1.50 or $2 a pair. These economics were observed at the time when Agnes MacPhail, a socialist member of Parliament, rose each year and vehemently and, I suppose, conscientiously argued that the Defence budget be one dollar *per annum*.

In the summer of 1938, the practical portion of the Militia Staff Course was held in Port Hope, where the facilities of Trinity College School were used to house and feed the candidates. For some reason, the regular force lacked sufficient instructors for the large number of Militia officers in attendance and asked a few of us who had taken the Advanced Staff Course to act as instructors. I was one of these, and through acting as an instructor I benefited greatly from the discussions with such excellent permanent force officers as Colonel Tommy Burns, later lieutenant-general in command of the United Nations forces in Egypt.

In the spring of 1939, King George VI and Queen Elizabeth toured Canada. They travelled by train throughout their visit. On Sunday, May 21, they were scheduled to stop in Kingston at about 3:30 p.m. on their way to Toronto. The plan, I am told, called for about an eight-mile drive around the city, with stops at the City Hall and the Royal Military College. Militia units from Military District No. 3 were ordered to line part of the route, and my regiment, about 150 strong, boarded the special train that had started from Bowmanville, about sixty miles west of Trenton, and picked up troops at the several towns from there to Kingston. As was the custom, we were in our assigned section of the route shortly after noon and had been told that the procession would be along about four o'clock. There were no great crowds along our section of the route because most of the fifty thousand people who had thronged into the city from communities east, west, and north of Kingston had assembled in the centre of the city or on the slopes of Fort Henry, where it was rightly assumed the royal party would drive slowly through. We waited and broke off to get a Coke or an ice-cream cone, and waited and broke off to answer the calls of nature, and waited some more. Finally, when the light was beginning to fade, Their Majesties and escort appeared and departed down the road before we scarcely had time to salute! I am told that the crowds at Fort Henry and in the city centre could scarcely see Their Majesties in the fading light. What a pity. Why the lateness of their arrival in Kingston? Prime Minister Mackenzie King decided that one or two unscheduled stops should be made between Ottawa and Kingston. He did not think of the thousands of people, many with families of small children, who were waiting for hours in the streets of Kingston. I recalled this episode twenty years later, when I was much involved with a royal visit!

7

BATTALION AND BRIGADE COMMANDER

I t is difficult to find words to express my feelings after I replaced the phone receiver that Friday afternoon of August 31, 1939. Colonel Young informed me that the regiment was being mobilized. When I had enlisted in 1916, it was with excitement and enthusiasm; the sense of adventure was in the air. I was leaving my family, but they were not dependent on me. I had no business to abandon. I was young and the future was mine. There were new countries to see, new friends to be made. Now, twenty-three years later, I was in my forty-second year and all was different. I hated the very thought of leaving my wife and son of six years and the home we had made and enjoyed together. Leaving a legal practice I had worked hard to develop and facing the thought of a substantial financial loss and going abroad now held no attraction for me.

What a fool I had been to spend time and money these past seventeen years on military affairs! Yet, I thought, I must be honest. I was only one of hundreds of Militia officers with families across Canada who would be feeling the same unhappy situation. We were not under any legal obligation to serve outside Canada until we signed an undertaking to do so. But to refuse to do that would have been in my opinion morally wrong and disloyal to the regiment and to the establishment that had trained me to carry out the duties I would now be required to perform. About these matters Jean agreed with me, though with sadness and reluctance. And so it was that I signed up for overseas service. Allan McNab and Isabel Hinds carried on the law practice. I went to war.

Busy days followed. Recruits appeared at first in great numbers. We had only World War I uniforms, old Ross rifles with bayonets, and no other weapons. Indeed, it was weeks and even months before all recruits could be properly outfitted from head to toe. And, sad to say, it was soon evident that my contention, written in the papers I had submitted on the Advanced Staff Course, was correct. The attitude of the people of Canada was such that a large army of four or five divisions in the field could not survive unless conscription was introduced at the very onset of hostilities. There was little of the enthusiasm for "the cause" in 1939 that there had been in 1914. No parson was dismissing his congregation after only a few chosen words so that they might attend a recruiting meeting in a local opera house. The result of this lukewarm or even antagonistic approach to supporting any war effort was quickly felt in our regimental area. The first spate of volunteers soon dried up, and we were hard put to reach our established strength of almost one thousand. However, by means of advertising and holding meetings in neighbouring counties, we did by near the end of October attain our objective and the whole regiment for the first time was assembled in Picton. Here the troops were accommodated in the basement of the armoury and in an old canning factory made habitable by installing makeshift heating, washing, and latrine facilities. The next six weeks were spent in "basic" military training at the nearby Picton Fairground and in route marches and other exercises to toughen up the troops.

We were frequently visited by officers from District Headquarters in Kingston, and I must acknowledge that they were most helpful in advising on training programs and in getting supplies and equipment to meet our needs. One day in late November, we received word that Major-General McNaughton, the General Officer Commanding the 1st Canadian Division, would arrive from Ottawa at Picton Fairground at 1400 hours to meet and inspect the regiment. On the appointed day, which was bitterly cold but sunny, we were duly mustered, more than nine hundred strong, and made a brave showing at about 1330 hours. We must not be late to greet the General! 1400 hours arrived but no General. The troops were getting very cold in a bitter wind, so the commanding officer had them march around for a few minutes and then form up again, expecting the General at any minute. This sort of activity went on all afternoon – march around, form up, march around, form up. At last, at

1600 hours, the General arrived. The compliments were paid, the inspection completed just before dark. It was a day long to be remembered by the troops, and the General was terribly embarrassed. What had happened? His staff officer was new to the job, and when he was told by the General to arrange for his visit for 4:00 p.m., he had mistranslated it to be 1400 hours and had so advised us, instead of converting it to 1600 hours!

New weapons and other equipment and the new battle-dress clothing were received. All were made ready for a move overseas in the near future. And the "near future" materialized in mid-December, just a week before Christmas. The commanding officer, Lieutenant-Colonel Sherman Young, a United Empire Loyalist descendant and veteran of World War I, was sent to Halifax to assume the role of commander of all troops that would cross the Atlantic in the P&O Line's s.s. *Ormonde,* one of a large convoy that would transport the 1st Canadian Division to England. On the departure of the CO from Picton, I was left in command to organize and control the move of the regiment by two special trains from Picton to Halifax. All went according to plan, and in due course we embarked in the *Ormonde* and lay for a day or two in Bedford Basin, awaiting the loading of other ships and formation of the convoy.

The first night aboard, seven or eight senior officers had the pleasure of dining at the captain's table with Sir Charles Mathewson, the senior commodore of the P&O Line, who had been recalled from retirement at the outbreak of war. After that evening, we never saw him again, as he spent his entire time on the bridge or a few steps from it. He was a charming old gentleman and informed us that this ship had never before crossed the North Atlantic and had always sailed in tropical routes from Britain to the Orient. Hence, he was sorry to have to tell us, there was a fine cooling system with plenty of ceiling and other fans but no central heating! However, he had had a number of electric heaters put aboard, which might give us some relief from the cold weather we would experience during the next fortnight. How right he was! We sailed on December 22, and what a dismal, bleak, and frigid voyage it was. Our course took us almost at once into Arctic waters, skirting Greenland and Iceland, and we were struck by storms of blinding snow and sleet until we got near the Gulf Stream. The troops were fairly warm below decks, and few of them ventured "top side" until we were nearing British waters. In the officer's lounge we had several of the captain's electric heaters and, with the help of these and wearing our great coats, we whiled away the hours, reading or playing bridge or poker for small stakes.

Our convoy of about nine transports had a naval escort of three Canadian destroyers, sometimes visible but usually ranging near the horizon, plus two French warships, the newly commissioned battleship *Dunkerque* and the cruiser *Gloire*, the former far off to the southeast and the latter away to the southwest. Leading the centre line of transports was the British battle cruiser *Renown*. A day or two after sailing, the Canadian destroyers returned to Halifax and we were picked up by three British Royal Navy ships as we neared Ireland. I seldom ventured on deck, particularly after dusk, but to a landlubber it was quite frightening to see how closely at times the ships came to one another, especially after nightfall when not a glimmer of light was showing and I was told that often the skipper of one ship had to be guided by the white water in the wake of ships ahead or abreast of him. The *Renown* seemed to lay very low in the water, and at many times in stormy weather the great waves broke over her foredeck and almost reached the bridge.

Christmas Day was a day of sadness and thoughts of home, accentuated, I suppose, for some of us, when on that morning we unwrapped the gifts that had been lovingly wrapped and stored away in our steamer trunks. Officers had trunks in which to pack uniforms and spare clothing. There was little to laugh about on this voyage, but one inconvenience was not without its humour. The officer commanding all troops on board – there was an Army Service Corps unit from Calgary and some other small units on the *Ormonde* in addition to the Hastings & Prince Edward Regiment – was Lieutenant-Colonel Young. He needed an office, referred to as an orderly room in the army, so a desk and a few chairs were put in the nursery for his use. Its walls were decorated with almost life-sized pink elephants, red monkeys climbing trees, colourful giraffes and zebras, and other brightly coloured birds and beasts, all cavorting across the walls under a brilliant and broadly smiling sun. Here the Colonel held court!

On the evening of the ninth day, we sailed into the estuary of the Clyde, and early on January 1, 1940, tied up at Gourock. It was past midday when we finally disembarked and boarded our train for an overnight trip to Aldershot Camp, south of London. My thoughts went back a little more than twenty-three years to the day I disembarked from the s.s. *Northland* at Liverpool and travelled overnight in exactly the same kind of train with the same type of rations, bread and bully beef with hot tea, cold coaches with a very dim bluish light and tightly drawn curtains, to Witley Camp, just a few miles from my present destination. History, indeed, was repeating itself.

About 9:00 a.m. January 2, 1940, we pulled into Aldershot and were met by the Brigadier commanding the 1st Canadian Infantry Brigade (1.C.I.B.), Armand Smith from Winona, Ontario, and Major-General McNaughton, the divisional commander, whom we had last seen on the Picton Fairground. With the Royal Canadian Regiment and the 48th Highlanders of Toronto, the Hastings and Prince Edward Regiment formed the 1st Canadian Infantry Brigade. It was a short march from the station to our quarters in Maida Barracks. Much has been said and written about that winter in England. It was the coldest in sixty years, and certainly it took its toll of Canadian soldiers who had been accustomed to central or at least adequate heating in their homes. Maida Barracks had been built at the turn of the century. There were no stoves except in the kitchens. Tiny fireplaces, for which damp soft coal was provided, were quite useless in keeping the thin-blooded Canadian soldiers warm and healthy. The CO went to Aldershot town and bought the few stoves that were available. Orderlies were detailed to keep the little fireplaces burning throughout the night, but in spite of all these efforts a third of the regiment came down with colds and more serious ailments. I was ordered to bed by the medical officer for ten days.

Toward the end of January, King George VI came from London to visit informally the troops. There was no great ceremonial parade. The troops lined the roads in their regimental area, and His Majesty simply walked along, stopping to say a few words with individual soldiers. I was temporary acting commanding officer and met the King at the border of our area and escorted him through our lines, followed by Brigade Commander Armand Smith and Major-General McNaughton.

Early in February Lieutenant-Colonel Young was returned to Canada; he had served in World War I with the 2nd Canadian Battalion and in 1915 at St. Julien was gassed and badly wounded and taken prisoner. Sherman was a brave and efficient soldier but like a number of other commanding officers who took their units to England in 1939-40, he was found to be too old and physically unfit for command.

Our new commanding officer, replacing Lieutenant-Colonel Young, was Lieutenant-Colonel Harry Salmon, a permanent force officer who had been a staff officer in Kingston in the 1930s and knew our regiment and many of us well. Harry was also a World War I veteran, had been twice wounded, and had been awarded a well-earned Military Cross. Some of our officers were a bit miffed because I had been passed over for command, but I was quite happy about having Harry as my co and we got along tremendously well together. One reason for being happy at not having command at that time was that there were a number of officers who were not up to the standard of training required to prepare men for battle. They and their families were long-time friends of mine, and it would have been painful and difficult for me to do what needed to be done. I felt that Salmon, with his professional knowledge and years of experience, would do better things for the regiment than I could do at that time, and furthermore he could and would teach me valuable lessons on training and conditioning troops for battle. The welfare of my regiment was of much greater importance to me than a promotion in rank.

Salmon's hand was soon felt. A number of officers left, new ones arrived. One of these was Major Ralph Crowe, a permanent force officer who came from the Royal Canadian Regiment to command one of our companies. Training programs were arduous and lengthy. At the end of each day the co had a conference in his office and discussed what he had seen during the day as he went to watch each company doing its particular task. He was not above giving a word of commendation where it was deserved, nor was he short on criticism as to training and administration. These meetings dragged on sometimes, so that we who attended them got to our dinners after 7:00 p.m., having been out since early morning. Captain Reginald Abraham, who had been customs officer in Picton, had a good sense of humour. At one of these long discourses, the co was interrupted by a distinct tinkling sound from Abraham's direction. The co paused, looked at Abraham, and, as the tinkling continued, said, "What's that?"

"Oh," said Reg, reaching for his blouse pocket. "Sorry, sir. I guess that's my alarm watch. I had it set for seven o'clock." Old Harry saw the point, smiled, and said, "I didn't realize it was so late. That's all, gentlemen." He was a bit more considerate thereafter.

In the early spring, during the "phony war," when we were still getting bacon and eggs from Denmark for breakfast, the senior officers in the Aldershot area were gathered in the large theatre to hear Field Marshal Sir Edmund Ironside, the Chief of the Imperial General Staff, who had come from London to deliver a lecture on the future conduct of hostilities. The Field Marshal was all optimism. One reason he gave for this outlook was, to use his own words, "You know, gentlemen, the Germans haven't got a chance. Hitler was a corporal and they haven't a general who was above the rank of captain in the last war."

As Salmon and I left the lecture hall, I said, "My God, Harry, did you ever hear such nonsense? And he is the senior officer of all British forces!"

Harry replied, "It was a stupid speech and I am surprised that he hasn't been replaced before this." In fact, Ironside was replaced a short time afterward, but the idea that World War II was going to be a repetition of World War I persisted in the minds of the Allied military hierarchy for a few more months.

Accordingly, the 1st Brigade received orders to train in trench warfare on Salisbury Plain. We were loading equipment on the train at Aldershot for the trip to Salisbury on May 11 when word came through to "stand-down" and await further orders. The Germans had launched their attack on the Western Front a few hours before, on May 10, and apparently higher command decided to cancel or delay departure of our brigade on the chance that we might be needed in France. However, the stand-down was cancelled two or three hours later and off we went to Salisbury. Here we dug trenches, sand-bagged them, lived in them day and night, "stood-to" at dawn and at sunset, and carried out mock raids on another regiment of the brigade across a hundred yards of "no man's land." In fact, I thought I was back in front of Arras with the 15th Battalion in 1917!

Then, after about ten days, we were suddenly ordered back from the trenches to our tented base camp near Tilshead, a small town on the Plain. There we beheld on the side of the hill beside the camp the greatest collection of road transport I had ever seen, everything from tiny Austin cars to large trucks and brightly coloured buses. These vehicles were to take us back to Aldershot post-haste. Quickly we packed up and in a great conglomerate convoy off we started through the lovely English countryside. Being second-in-command of the regiment, I rode at the rear of the line in a small car and had rather a slow journey because a number of vehicles broke down and it was my job to arrange replacement and transfer of loads onto spare vehicles in the rear of the convoy. So it was that by May 20 we were back in our old Maida Barracks. The Germans were advancing rapidly, having forgotten about the trenches of the First War!

Scarcely had we unpacked our gear, wondering what would happen next, when we were put on two hours' notice to move again, this time to

France. One can scarcely conceive the confusion, the orders, the counter-mands, and the uncertainty as to what would happen on the morrow. We, in a regiment, were small pawns in a gigantic game. Churchill was now Prime Minister and brought some semblance of hope and determination to all of us, but he and his colleagues were faced with problems of unbelievable peril and complexity. The 1st Canadian Division was the only complete, fully equipped formation left in Britain. Should it be thrown into the fray in France or kept in Britain as a last mobile, trained but untried force in the event that the Allied armies on the continent collapsed, as it appeared they would? General McNaughton was sent by destroyer to Calais and Dunkirk, still in Allied hands, to view the situation and bring back to the War Office his opinion and advice. The 1st Brigade moved by trains to Dover to be ready for the crossing to Calais. The 48th Highlanders went aboard ship, the Hastings and the RCR were at dockside ready to do likewise. This was the situation on May 24, 1940. Then came the result of McNaughton's reconnaissance and opinion: it would be folly to commit the Division to the intended landing and defence of Calais as they would be destroyed or captured. And so back we went to Aldershot once more. We never detrained at Dover, and the episode was known to me and others involved as "the Dover Dash."

Within the next few days the Germans had driven the wedge between the French and British armies on the continent, and the evacuation through Dunkirk began. On May 31, the 1st Canadian Division was ordered to move from Aldershot once more, this time to the area of Northampton, north of London. I think there were two reasons for this. One was the need for our barracks in the Aldershot area for troops coming from Dunkirk. Another, of more importance, was that we would be in a position to counterattack any enemy landing on the beaches of East Anglia between the Wash Estuary and the Thames. The move to the north of London was our first by road with our own vehicles and it took place at night. All road signs had been removed and the regiment, headed by the CO, Lieutenant-Colonel Salmon in his station wagon, had to find its way to its destination, the small town of Finedon, near Northampton, with the aid only of one-inch ordinance maps and a few personnel who were sent ahead in daylight to act as traffic guides at certain road junctions. We moved without lights! At one point we found ourselves moving in a southerly direction instead of northerly, but we were able to make a circuit in a farmer's field. By mid-morning we were at Finedon.

Here the troops were billetted in factories, barns, houses, sheds – any place for shelter. I was fortunate in finding a very good room in a doctor's home, the best billet I had had since leaving home!

We immediately did a reconnaissance of roads leading to the beaches of the North Sea, between the Wash and the Thames, and practised movement both by day and by night to positions from which we would hold and repel (hopefully) the invaders, should they come this way instead of the shorter route across the English Channel. The old town of Finedon had little rest, nor did we, for the next week. Buglers on vehicles at all hours raced through the streets sounding "the alarm" or "reveille." Security was tight; everyone was suspect. One battalion commander of our brigade spent a night in a jail near the coast after being challenged by a local defence volunteer. The commander did not know the password.

After a week of this life on tenterhooks, we were ordered to return to Aldershot at once. This move was by daylight and there were no detours through barnyards and pasture fields, but one of our vehicles did knock the corner off a building in the town of Nettlebed. There were no casualties, fortunately, and little damage to the truck. Worse than that was the damage done to our armoured personnel carriers, which we had received just before the move to Finedon. They were driven at the same speed as the rest of the convoy, without a stop from Finedon to Aldershot, and as a result the rubber on their bogey wheels was burned off. Carriers were not meant for that kind of a trip over hard-surface roads. The result was that for the next few days they were at the ordinance depot in Aldershot where the mechanics worked day and night, with me pushing them, to repair the carriers. Once again we were ready to move. And move we did.

On June 9, orders came to send our motor transport to Falmouth by road for shipping to France, and we were placed on one hour's notice to move. Two days later we were ordered to prepare for movement by train the next evening, destination Plymouth. In late afternoon of June 12, we once again marched from Maida Barracks to the Aldershot railway station, each man heavily loaded with equipment, weapons, extra ammunition, and rations consisting of loaves of bread in sacks, tins of corned beef, and tea. The only vehicles with us were bicycles, and they were draped mostly with sacks of bread. I cannot say that there was any great excitement among the troops. They had been "bitched about" a great deal during the past month. Most of us realized that an entry into

France, after the French debacle and the evacuation of British forces from the beaches of Dunkirk, would be a most perilous venture.

Our overnight train took us from Aldershot to Plymouth, where our railway coaches were spotted for detraining at the dockside. Movement central officers showed the co and me our ship, which was to take us to Brest, at the tip of the Brittany peninsula. She was a fine-looking, French coastal liner called the *Ville d'Alger* but we could not yet go aboard because she had brought from Brest a cargo of strawberries, which had not yet been put ashore. This was more than a week after the last British soldier had come from Dunkirk, and yet a great white ship had sailed from Brittany to Britain with a cargo of strawberries! It was late afternoon before we got the troops aboard, they having downed many quarts of very tasty fruit, which was within easy reach of the licentious soldiery.

The overnight voyage was nerve-wracking but proved uneventful, and early next morning we berthed at Brest. Just after sunrise I came down from the boat deck where I had spent the night and was reading the notices on the bulletin board at the purser's office. The co came along. "Did you get any sleep, Howard?" he inquired.

"Yes, a few hours, but look at this," I replied, and read him one message in French over the printed signature of Marshal Petain. I wish I had made an exact copy of this document, but its words were clearly to the effect that it was impossible for France to continue the struggle and further fighting would be suicide. I said to the co, "It seems to me that the sooner we get back to Plymouth the better it will be."

"Oh, I'm afraid we can't do that, Howard. I agree it looks bad, but orders must be obeyed," replied the co.

On June 14, a very hot summer day, we disembarked from the *Ville d'Alger,* and under instructions from the British movement control officers who met us at dockside, we formed up the regiment, as for a route march. Still with our bicycles and heavily laden equipment, we made the mile-or-so trip up a steep and very dusty hill to an equally dusty, dirty park, which resembled a park only because of the presence of a few scattered trees. Here we rested and awaited further orders. Along the road in front of the park small groups of French soldiers wandered toward Brest, mostly unshaven and unkempt, but seemingly happy and in good health. When they saw us they would laugh, wave their hands in a derisive gesture, as much as to say, "What do you think you jackasses are going to do?" Their appearance and attitude reinforced my conviction that the sooner we got out of France the better it would be. About 4:00 p.m., the movement control officer came along and told the co that a train was now ready and would take us forward (inland) to the area of Le Mans, and there we would get further orders. From the park we marched down to the railway yard, and there indeed was a passenger train, which took the whole regiment aboard and off we started.

"Started" is a good word because we would move a few miles, then pull into a siding to allow a trainload of refugees to pass on their way to the coast. At each station where we stopped the platform would be crowded with men, women, and children fleeing to they knew not where, but frightened to death of the dreadful Hun. Where possible at each station stop, the co and I would walk back along the train to see that all was well and that the officers had the men well in hand and were not allowing any to leave the train. On one such amble who should we meet but our paymaster lugging an armful of bottles of white wine he had bought at a shop in the station. I thought the co would explode as he ordered him to abandon his bottles, which he did. Through the night of June 14-15 we made our way slowly toward the east, and there was neither sight nor sound of enemy aircraft. A few bombs or machine guns could have made mincemeat of us. Near dawn, we stood stalled on a high trestle bridge. I thought, "Boy, oh boy, what a trap. If we are attacked by air, there is no place to go except a long jump into the ravine!"

The next day was warm and bright. Our train continued on its apparently uncertain way for a few hours and then stopped at a small station and went no farther. A British officer appeared and said we were to return to Brest. After a short pause, we started to back up and we travelled in reverse, with engine still at the head of the train, the route we had taken the night before. It is difficult to describe the scenes of confusion and crowding at all the stations we passed through. Many of our men who had not eaten all their iron rations or had acquired some loaves of French bread passed them out to women and children on the platforms. We stopped at the stations, presumably for the crew to get further orders. Our train had run short of water, and the co relented to a point where he acquiesced in the purchase of wine when it was within reach. Many a water bottle was filled for the first time with something other than water! Eventually, late on the fifteenth, we returned to dockside and there found the British cross-Channel steamer the *Canterbury Belle* tied up and apparently ready to take us from Brest to

Plymouth. The Royal Canadian Regiment, which had been through an exercise similar to ours, was already aboard, and we learned that the 48th Highlanders, the third regiment of the 1st Canadian Brigade, had been taken to St. Malo and ultimately reached home at Aldershot before we did. The troopers referred to this experience as the Bust at Brest.

The *Canterbury Belle* remained tied up at Brest until near dusk the next day. The little ship seemed loaded to the limit, but there still came aboard occasional British personnel – Salvation Army canteen workers, a brigadier alone with only a small haversack on his shoulder, who had escaped when his headquarters was overrun. At night we slept or dozed, lying down if there was room, or sitting and leaning against a comrade. As for myself, I found a spot under a table in the dining saloon with men sleeping on top of it. With three other card sharks on Sunday, I played bridge on the top deck under the shelter of a lifeboat. In mid-afternoon we saw our first and only aircraft, a German recce plane, the pilot of which either had no ammo or had a merciful heart because not a shot was fired at the *Canterbury*. For the few seconds that the plane was near enough, a terrific fusillade of fire went aloft from bren guns, rifles, and even revolvers in the hands of the troops on board. For the rest of the afternoon we surely expected the worst, on the assumption that the pilot of the recce plane would report our position to his masters and a massive attack would result. But all remained peaceful and we continued our game of bridge! Perhaps the victorious enemy were all at church parade!

At sunset we cast off and early the next morning berthed at Plymouth, having had a safe and uneventful crossing. I never did learn the reason for leaving us tied to a dock in Brest for twenty-four hours, but perhaps there was no escort available. Perhaps the War Office and Prime Minister Churchill thought there was still, even as late as June 16, a chance that the French might be willing and able to make a stand and, if so, we might still be able to assist them. Whatever the reason, it was a frightful, dreadful decision to leave two thousand men at the mercy of an enemy which, by the use of half a dozen aircraft, could have annihilated us.

At Plymouth, after an absence of only four days, we were greeted by the mayoress, Lady Astor, and a bevy of assistants with hampers of sandwiches made from bread and bully beef. I was with the CO when he and William Hodgson, the CO of the Royal Canadian Regiment, discussed which unit should be first to disembark and entrain for Aldershot. "Uncle Bill" was a great gentleman, perhaps a year or two older than

Harry Salmon, his regiment was senior to ours, and he could certainly have said, "No dice, Harry, you and your Plough Jockeys will have to follow us ashore and we'll meet you at Aldershot." However, instead of that, he said, "Okay, Harry, let's toss a coin for it." They did and Uncle Bill lost, so we, the Plough Jockeys, came ashore and were back home in Aldershot a few hours before our good friends in the Royal Canadian Regiment.

But not all of us were yet home. Our vehicles and drivers had arrived in Brest a day ahead of the regiment. They had been directed eastward and told that we (the rest of the unit) would join them in a day or two. The plan, I learned later, was for the whole 1st Canadian Division to come to France and join hands with the 51st Highland Division and a number of French formations to form a defence line across the base of the Brittany peninsula. Of course this became impossible when, before we could arrive, the enemy surrounded and captured the 51st Division and the French collapsed. Our vehicles were eventually stopped, as our train had been, and ordered back to Brest. But on arriving near there, after we had sailed in the *Canterbury Belle*, the drivers, with the vehicles of the other two battalions, found no ships available to carry them to Britain. Orders were given to destroy the entire lot of trucks, and this was done by slashing tires, cracking engine blocks, and burning the chassis as far as possible. The drivers, by good fortune and with the help of the kindly French peasants, had food and water (and wine) enough to suffice for a few days until a British destroyer came in and carried them across to Portsmouth. Thus, this whole operation, which might have been a dreadful disaster, resulted, in the final analysis, in the loss of three men (captured) and our entire establishment of some eighty vehicles. On balance, we were lucky.

For a few weeks we were grounded in Aldershot until a new lot of trucks and carriers arrived. Then for the last time we left Aldershot and moved south into Surrey, to spend the rest of the summer in bivouacs and tents in the countryside, each company having its own area of fields and woods in which to live and hide their vehicles. It was during this period that we learned that Prime Minister Churchill was coming for a day to see the Canadians but would be reviewing only the 2nd Brigade, which was commanded by Brigadier George Pearkes. Thereupon hangs an amusing tale often told by Pearkes himself. It seems that the Prime Minister was to come from London by car and would meet Major-General McNaughton and Brigadier Pearkes at a certain well-known pub at a crossroad, identified by a map reference because there were no signposts on the roads. The day dawned dark and rainy and it continued to rain heavily all day. The General and Brigadier Pearkes and their principal staff officers were at the designated pub at the appointed time. I think it was 10:30 a.m., but no PM appeared. After a twenty-minute wait, they decided that the PM might have made a mistake and gone to another pub a mile away at another crossroad, so they sent the brigade major of the 2nd Brigade, Rod Kellar (later Major-General Kellar), to find the distinguished visitor. And sure enough he found him at the other inn, comfortably seated before a blazing fire, smoking a cigar with a libation at his elbow. Rod smartly saluted the Great Man and said, "Major Kellar, of the 2nd. I am afraid you have come to the wrong place, sir. General McNaughton and Brigadier Pearkes are waiting at the Pig & Whistle down the road, sir."

Churchill replied, "Not at all, Major. I have made no mistake. I know how to read a map. Tell General McNaughton I am waiting here."

Rod quickly responded, "Yes, sir," and departed. Shortly thereafter the party were assembled at Mr. Churchill's inn!

This was only the start of a dismal day that produced a comedy of errors. After a snack and a bit of visitation, they went forth to see the brigade at work. Brigadier Pearkes explained that they would first see a regiment practising an assault on an enemy position. When they got to the appointed area, the exercise had been finished and the unit had departed to their tents to dry out. Too bad. Next the Brigadier said they would see a regiment with all its transport move along a road where the PM could take a salute. So the party took up their positions at the appointed place, and right on time the lead vehicle hove in sight half a mile down the road, but, as Rod later described it, "Damned if they didn't turn off at an intersecting road about 300 yards from where we were." The PM never did get the salute. Too bad again. Then the party went to see the way the troops were "making do" in their tents and bivouacs, which in the rain was not a particularly enjoyable event. And so the day of miscues ended, but not quite! About 7:00 p.m., Brigadier Pearkes had a phone call from Division Headquarters asking if he was not coming to the dinner. He had never heard of the dinner. He had been overlooked and not invited! George Pearkes would laugh in later days when he told of the Churchill visit, but at the time he was mortified.

Early in August, 1940, we were training with tanks. Salmon was called to Division HQ. I carried on for the hour or two he was away. On his return he took me by the arm and we walked away along a Sussex lane. "Howard," he said, "I have had news. I have been posted to Second Division as General Staff Officer (GSOI), and Major Edgar of the Patricias has been appointed to command the regiment. I'm terribly sorry you are being passed over again. You deserve the command and I am sure your turn will come." I was stunned and terribly hurt. When Salmon, a regular officer, replaced Young, I was not unhappy. We knew each other and I had great respect for him and his ability to do the things that I would have found very difficult and embarrassing to do. I had loyally supported him during these six months, always assuming that when he went I would succeed him, and this was his wish and recommendation. The next day, the newly made Lieutenant-Colonel Edgar, formerly second-in-command of the Princess Patricia's Canadian Light Infantry, arrived. No one knew him and he knew no one in our regiment. He was a permanent force officer, as Salmon had been, and he had a kindly, gentle disposition. He spoke with a pronounced English accent, carried a walking stick, and (a major *faux pas*) arrived at our unit wearing PPCLI badges and their special type of head dress. When Salmon had arrived, he was dressed as a member of our unit and not wearing the badges of the RCR, which had been his regular unit.

Jimmy Edgar was with us for four weeks. I do not want to be unkind, but he was not meant to be a regimental commander, least of all in command of the Plough Jockeys. He departed and returned to Canada. On September 2, I was promoted to the rank of lieutenant-colonel and appointed CO of my old unit. I was mollified and happy.

The regiment met with a veritable spate of misfortunes within a few weeks of my taking command, and none was of my making! First, a civilian on a motorcycle, travelling at high speed, ran squarely into the front of one of our trucks and was instantly killed. A court of inquiry had to be convened to establish the facts. This was a nuisance. Second, the regimental water-tank truck, in avoiding a civilian vehicle, ran off the narrow road into a ditch, and one of our men, riding on the fender (where he should not have been), was killed. Another court of inquiry was convened. Third, two subalterns one dark evening took a small truck, called an eight-hundred-weight (8 CWT), to a pub in Dorking for a few beers (which they should not have done). On leaving the pub they backed from their parking place and, in doing so, they heard a "thump." On getting out of the truck, they believed they had bumped into a brick wall. Instead, they found the body of a dwarf whose head had been crushed between the tail-board of the truck, about three and a half feet above the road surface, and the wall. I gave the officers some credit for not driving off in the dark and leaving the poor victim. They carried him into the pub, waited until a doctor and the police arrived, explained what had happened, and were absolved of blame by the civil authorities. But I had to deal with them to the limit of my powers as a CO – extra duties, no leave, and severe reprimand. They were good officers and toed the line thereafter. Fourth, there was the invader. "A" Company were in bivouacs, with some bell tents, around the edge of a large ten-acre meadow bordered by trees and shrubbery. This was in September, 1940, when fear of an invasion was in everyone's mind. Passwords were frequently changed so that if anyone could not respond properly to a challenge, he was suspected of being an enemy parachutist. On a quiet, very dark night, a sentry at "A" Company with loaded rifle heard a distinct rustling in the shrubbery and called the challenge and got no answer. All was quiet again for a few minutes, then the rustling again, and again no reply to the challenge. The sentry was taking no chances. He raised his rifle and fired at the rustler. He heard a crash and all was quiet. This had wakened the sergeant, of course, who searched and found the victim, a horse, which was said by its owner to be the best horse in the county, worth a great number of pounds sterling. The soldier who had fired such a remarkable shot in the dark was not held responsible, and after much negotiating the farmer received from Canadian military headquarters about five hundred pounds! That was the final nuisance.

The great German air offensive was launched against Britain on August 8, 1940, and continued with unabated fury until mid-September. But by then the Battle of Britain had been won. The German attempt to neutralize our air defences had failed. On a glorious warm and sunny Saturday, September 7, we received the warning word "Cromwell," meaning an invasion was imminent and all units must be prepared to leave on four hours' notice to take up defensive positions by routes we had reconnoitered and could follow in the darkness of night. This was only five days after I had assumed command. I was terribly nervous but hoped that I did not show it, as I called my "O" Group (company commanders) together and issued the orders for the expected move to the south. Afterward I visited each company area to be sure that all was ready. This state of preparedness continued for more than a week. On the following Wednesday, Mr. Churchill broadcast to the nation thus:

If this invasion is going to be tried at all, it does not seem that it can be long delayed. Therefore we must regard the next week or so as a very important period in our history. It ranks with the days when the Spanish Armada was approaching the Channel and Drake was finishing his game of bowls; or when Nelson stood between us and Napoleon's Grand Army at Boulogne. We have read all about this in the history books but what is happening now is on a far greater scale, and of far more consequence to the life and future of the world and its civilization than those brave old days of the past.

By Friday, September 13, it was estimated, by reading air photographs, that boats and ships capable of moving 175,000 men and their equipment across the Channel were gathered in enemy ports from Holland to Brest. Two days later, on another clear, warm, September Sunday, the most intense air battle raged over the south of England, and in our regimental area we picked up three German pilots and one British airman who had parachuted to safety from their flaming craft. It was reported that on that day 185 German planes were shot down with a loss of only twenty-five British. I have always doubted the accuracy of this report, but nevertheless, Goering's Luftwaffe had been badly mauled.

From that day onward attacks were few and light, and within a week the concentration of enemy shipping was moving away from Channel ports. The threat of immediate invasion was over, but continued vigilance and defence against possible enemy raids on the British coast were still essential. I, for one, heaved a great sigh of relief when we received word that we could stand-down from our fortnight of almost instant readiness to move to battle positions. My first two weeks as a commanding officer had not been without tragedies, tension, and excitement. I was fortunate, indeed, to be with my own old regiment, where I had officers, non-commissioned officers, and men, most of whom I had known a long time and in whom I had great and justifiable confidence. My regimental sergeant-major, Angus Duffy, had been a neighbour in Trenton. My second-in-command, Major Bruce Sutcliffe, several years younger than I, had been at public school when I was in high school, and he was a fine soldier.

In the summer and early fall of 1940, the centre of our regimental area was in and around Lyne House, an estate owned by the Broadwood family. Harry Salmon and, for his short term, Jimmy Edgar had kept a bedroom in the house, but I always slept in a bell tent, pitched nearby, which I shared with the padre, Captain Walter Gilling (later to be Dean of St. James Cathedral, Toronto). When October arrived, we were told by HQ to find winter accommodations in a listed number of large homes in that area, north of Horsham and south of a line running from Dorking to Redhill. Property owners knew that we had authority to requisition, and in most cases they were very helpful and co-operative in making large homes or parts of them available to all units of the 1st Canadian Division and many other British divisions as well. In mid-November I received orders to move the regiment to the south coast, about sixty miles away, and take over defensive positions from a British unit. This was to be for only a few weeks, so we left a small rear party in charge of our billets and returned to them at the end of December.

The tour of duty at the coast was interesting, but in no sense a sinecure. My headquarters was in a small barracks on the outskirts of Brighton, and my area of responsibility extended for some thirty miles along the coast and on the South Downs, which in that location overlooked the Channel. The troops were in billets, taken over from the British unit we had relieved, which included more than fifty small summer cottages or huts along the beaches at Brighton, a hotel, and outbuildings at the Devil's Dyke high up on the Downs, a school at Shoreham, and a factory at Lewes. The task of inspecting was time-consuming and difficult. To maintain a high state of morale in these circumstances required a continual and arduous program of training and a very high standard of administration and discipline. On Christmas Day, 1940, Padre Gilling, Sutcliffe, Abraham, and I set out in my station wagon from the barracks in Brighton before noon to visit all the companies on a prearranged schedule so that we would see them at their Christmas dinner. Turkey was on the menu. It was a dull but mild day, and we made the rounds as planned – the beaches, Shoreham, the Devil's Dyke, and the Downs behind Lewes. We returned at dusk to our barracks for our own Christmas feast, having covered almost seventy miles in visiting all the companies.

A few days later we turned over our positions to another Canadian unit and returned to our area between Dorking and Reigate, where we were destined to remain for almost a year. We were in Sussex, the home county of our allied British regiment, the Royal Sussex, and of course we saw much of them. Many of their men had been taken prisoner before the evacuation from Dunkirk, and their home depot at Chichester had set up a Prisoner of War Fund to provide parcels of food for their comrades in Germany through the facilities of the International Red Cross. I conceived the idea that members of our regiment, being fraternal comrades, so to speak, might make a contribution to this fund. My company commanders agreed, and the men seemed enthusiastic. So in each of our billets on a given pay-day was placed a box or basket, and all ranks were invited to put therein whatever they felt inclined to give. I am happy to say that more than £700 was collected. With the consent of our brigadier, we then invited the honorary colonel and senior Royal Sussex officers in the vicinity and our own divisional commander and brigade commander and other local VIPs to a formal regimental parade at Reigate. Here, on a lovely day in June, I presented a cheque to the Royal Sussex colonel for the amount collected, and, needless to say, he was most grateful. This I think was one of the "red-letter" days of that summer.

It was here, too, that I first made the acquaintance of "Monty," Lieutenant-General (as he was then) B.L. Montgomery, the commander of all troops in southeast England, where we were. Lieutenant-Colonel Harry Salmon, when commanding our unit, had set a high standard of training to make the troops tough and battleworthy. I tried to follow this policy. Monty even surpassed Salmon in his determination to make all ranks physically fit and mentally prepared to overcome the enemy. He preached his gospel at every opportunity. He planned and carried out frequent exercises involving real hardships, i.e., long marches, short rations, days and nights without sleep, so that the troops would experience as realistically as possible, in the south of England, what they must expect to endure in some theatre of future operations.

Monty frequently spoke before large groups of officers, usually in a theatre. When all were assembled, he would walk from the wings to centre stage and open his discourse by saying something like this: "Good afternoon, gentlemen. Before I talk about the exercise we are going to carry out next week, please clear your throats, get rid of the coughs, and blow your noses now. I don't want to be interrupted later." He would then wait a minute or two, while we all did what nature demanded, and thereafter there was absolute silence as the little man, without the aid of notes, very clearly and not wasting a single word, explained what the exercise would involve, which units would be the "attacking force" and which the "defenders," etc. On one occasion he stipulated that we should give the men a good breakfast each morning of the exercise because they would not have anything more until late that night and perhaps not even then. But Monty had a sense of humour. On one of these occasions he added, "Of course, I don't mind if the Canadians chew gum if they get hungry."

On two occasions Monty came to the 1st Canadian Brigade to see the units at their training. Both times the brigadier was ill, or he said he was, so I, as the senior battalion commander, had the pleasure of taking Montgomery around to see the three regiments. The first of his visits was on a miserable rainy day. At the first regiment we visited – I refrain from naming it – we met the commanding officer. Passing the time of day, Monty said, "Well, Colonel, what are the boys doing today?"

I almost fell off my chair when the co replied, "Excuse me, sir, I'll just call in my training officer." He did and the officer gave Monty a rundown on the day's training program, and then we saw the men at their various duties. The training was excellent, but the unforgivable fact was that the co did not know what his unit was supposed to be doing!

About noon we arrived at my unit, were met by Major Sutcliffe, my second-in-command, who was in command while I was taking care of Monty. All went well and eventually we came to "D" Company, commanded by Major Kennedy, who smartly saluted and said, "We have been doing battle drill this morning, sir, and the men have just broken off and are lining up for dinner. Would you like to see them, sir?"

"Yes, yes, I would indeed, Kennedy. Lead the way," replied Monty. The shed serving as a mess was close by and in we went. It being a rainy day, the men were wet, muddy, and dirty-looking. They were moving along with their mess tins and the steaming stew looked and smelled mighty good. Monty had a great time for fifteen or twenty minutes chatting and laughing with the boys. These Plough Jockeys, wet with rain and dirty with mud, were made to feel completely at ease in the presence of this great leader of men. I sat beside him in his car on our way back to brigade HQ for lunch, and he said, "You know, Graham, that's what I like about Canadians, not afraid of me. Our English officers would never do what Kennedy did, ask me to go and see his men get fed. Jolly good show, I enjoyed it."

On one of our exercises we moved into Hampshire and the area of southwest England where Lieutenant-General the Hon. Harold Alexander, later Field Marshal Earl Alexander of Tunis, was GOC, as Monty was in the southeast. I had heard much of Alexander: how he had been in charge of the evacuation through Dunkirk after Lord Gort had returned to England; how he had been the last British soldier to leave the beaches;

how he used to dance Irish jigs for the amusement of his men when he was a platoon commander with the Irish Guards in World War I; how when he was said to be the youngest battalion commander in the British Army he had won the Army championship for the mile run and was close to being a British representative at the Olympics in the early 1920s. In a word, I was much impressed by what I had heard and I hoped I might see him on this exercise, which I did. The General appeared one morning at my HQ under a tree, with an ADC. He wore his proper rank badges but otherwise was dressed as an officer of the Irish Guards. I saluted smartly and said, "Good morning, sir."

He responded, "Where are your companies located?" I had my map spread on a table with the company positions marked thereon and showed it to him. He glanced at the map for a minute or two, said, "Um, thank you," and walked away.

I was disenchanted, to say the least. Any other commander that I had met would have said something, asked a question, expressed criticism or agreement with my plan, shown some interest in the operation, but not so Sir Harold. I never met the General again during the war, but on several occasions we met at functions when he was Governor General of Canada, and later, in London, when I was stationed there, we would meet at Canada House receptions. Mr. Churchill had great confidence in him, and he left Canada to be in Churchill's cabinet. He was rewarded by a promotion in the peerage from Viscount to Earl. Others felt otherwise. Field Marshal Slim, certainly one of the great leaders in World War II, was once asked how Alexander had acted in Burma when Churchill sent him out there as commander-in-chief. Slim's reply was, "I don't believe he had the faintest idea of what was going on!" General Sir Ian Jacob, who knew something of his capabilities and limitations, said, "He never produced a single idea or suggestion during the entire time I served as his chief of staff." When Monty was asked about Alexander's capabilities, he said, "Well, you know, I taught him at the Staff College and we, the directing staff, came to the conclusion then that he had no brains – and we were right." But Alexander did have courage and inspired courage and loyalty in his staff and subordinates, in spite of his mental laziness or his lack of intellectual assets.

In the summer of 1941 my regiment located in the Dorking-Reigate area. The greatest danger we faced was that life was too easy and boredom, lack of discipline, and a lowering of morale would follow, despite Monty's desire to make training exercises "realistic." I was very proud of the way my Plough Jockeys did their part on these occasions, but also a bit embarrassed at times when they went overboard in their realistic capers. On one occasion, our "enemy" had light-tracked vehicles, and we were told to expect them to approach our position by a certain road. One of my company commanders, whose troops would first meet the enemy onslaught, discovered several drums of tar near the road that were to be used for road repairs. He and his men, treating the enemy in a realistic way, dumped the drums of tar on the road and then from ambush, as the enemy vehicles skidded and stalled in the awful goo, took the whole lot prisoners and we won the battle! But that left the enemy vehicles in an awful mess. Another time we were fighting a battalion of British Guards who had tanks, and we knew that on their line of expected advance there was a cutting in the road with banks ten-feet high on both sides for one hundred yards or so. Again, being realistic, my lads hid at the top of the embankment and, when the enemy appeared with their turrets open, threw shovels full of loose earth into the tanks on the road below them. Dirty pool! But effective. On a third occasion these former farmers, bricklayers, and rail splitters felled a good-sized tree that effectively blocked an enemy advance and thus provided an excellent opportunity for enfilade fire. Good marks came from the umpires but also a lecture by my divisional commander for destruction of property.

The fourth and last exercise showed particular realism. The enemy were holding the town of Horsham, and if we could get to their rear and cut their line of communication, we might be adjudged victors by the umpires. Time was of the essence, and to go around the town would take too long. So why not go through it? One of my company commanders stopped a double-decker bus short of the town, ordered all the passengers to the top deck, and put one of his platoons and himself on the main deck, told the driver not to race the vehicle but to drive at a reasonable speed straight through Horsham, which he did. The platoon debussed at the right time and sighted the automatic guns. The company commander got good marks for his ingenuity. These pranks may seem childish but they prepared all ranks of the regiment to use their heads as well as their hands and weapons in finding ways to overcome the real enemy when we eventually came to grips with them.

Our stay in that lovely area of Sussex was from the summer of 1940 to late in 1941. A goodly number of our men and eight or ten officers, as one might expect, married charming young ladies from Sussex. Many of them and their progeny now live in the counties of Hastings and Prince

Edward. A good friend of mine and of the regiment was J. Arthur Rank. He was a man with many business interests in London, film production being one of them, and had a very large house at Reigate overlooking the heath. Men from one of my companies were billeted in the servant wing of the house, and Rank and his wife frequently invited me and two other officers, who were also bridge players, to dinner on Saturday evening and for a few rubbers of bridge afterward for tuppence a hundred stakes. The Ranks also owned a large estate in Hampshire, where they had cattle and grew fruit and vegetables and had a large area to shoot over. As a result of these useful wartime assets, any guests at the Ranks' dinner table were treated to all sorts of tasty victuals, berries and cream, pheasant, melon, all beautifully prepared and served. Among other interests, Rank owned the Gaumont Graphic Film Company, and when we had the posh parade for the Royal Sussex, Arthur sent a filming team from London and they produced a splendid record of the proceedings that was shown in many of the theatres he controlled in Britain. He gave a print of the film to the regiment, but it was lost with most of our other memorabilia of great historic interest when the Picton Armoury was burned shortly after the war.

While still in Sussex, the 1st Canadian Division was visited by the Prime Minister, Mackenzie King. Before coming to see us, he had visited the 2nd Division, and the licentious soldiery had booed him, much to the embarrassment and chagrin of their officers. To prevent such a shameful performance on his visit to the 1st Brigade, our brigadier had warned the unit commanders that, if there was any mark of disrespect shown to our Prime Minister, there would be hell to pay and we would be fired. We put on a good and very respectful show, but the PM himself made an awful *faux pas*. After his inspection of my Plough Jockeys, I escorted him to the next unit, the senior infantry unit in the Canadian Army, the Royal Canadian Regiment. Lieutenant-Colonel Eric Snow, the CO, met us, but before either of us could say a word, Mr. King looked at Snow and said, "And what outfit is this?" It must be acknowledged that in the U.S. Army to refer to a unit as an "outfit" is quite acceptable, but with us the word had a bit of a derogatory connotation. The look on Snow's face clearly showed that he resented the appelation when applied to his very splendid regiment. He fairly shouted, "This, sir, is the Royal Canadian Regiment, ready for your inspection, sir." The ceremonies proceeded to a happy conclusion when we all met and chatted with the PM for tea at divisional headquarters.

All in all, the regiment enjoyed the long stay in the Sussex area, and it was with great regret that we, with the rest of our division, were moved to the south coast toward the end of 1941. So it was that once again my Christmas Day was spent, as it had been in 1940, within the sight and sound of breakers rolling along the beaches of the English Channel at Bognor Regis. Before the 1942 New Year, my battalion headquarters was located in the village of Middleton-on-Sea, between Bognor Regis and Littlehampton. As had been the case in the Dorking area, the troops were billeted in vacant homes, factories, huts, and schools. We kept busy with exercises (which took us for days at a time some distance from our base) but we never went so far but that we could man our coastal positions within a few hours should the need arise. We were also able to help local farmers in the type of work a good number of my boys were familiar with – ploughing, sowing, and reaping early crops of hay and grain.

A few incidents of the spring and summer of 1942 stand out in my memory. There was the day in February when I took the regiment on about a twenty-mile route march to an area on the South Downs, north of Chichester. Here my intention was to bivouac for the night, have an exercise early next morning with A & B companies pitted against C & D, and then do the return twenty miles to our billets late that day. Before going on a jaunt like this, it was necessary to inform the brigadier in detail where we would be, which I did. We had just reached the bivouac area when I had a message from the brigadier by dispatch rider, ordering me to return at once and with speed to our coastal position. An enemy raid was feared imminent. We did a forced march and in less than five hours were at our proper defensive posts. What had happened? It had been discovered that the three German battle cruisers, *Scharnhorst*, *Gnisenau*, and *Prince Eugene*, which since 1940 had lain in Brest harbour under the protection of shore-based German batteries and the watchful eyes of the Luftwaffe, had left their hideaway and were steaming toward the English Channel. For what purpose? Perhaps they were carrying or were intending to support a raiding force from the French-Belgian coast upon the Sussex coast. Or perhaps the three ships were simply making a move from Brest to a more easily protected anchorage in German waters at Kiel. That, in fact, is what they succeeded in doing, despite valiant, almost suicidal attacks by the slow-moving, ill-equipped planes of the Royal Air Force. Large ships of the Royal Navy were not nearly enough to engage the enemy, who had the audacity, in broad daylight and within sight of the British coast, to make this dash from Brest to Kiel with impunity. This occurrence was the only real scare we had that summer.

Our headquarters was located close to the town of Arundel and almost in the shadow of Arundel Castle, the home of the Duke of Norfolk. His Grace, Bernard by name, was wont to ride a bicycle around the countryside and was a friend to many Canadians. The Royal Canadian Regiment of our brigade had a company billeted on the Duke's estate, and one day in doing mortar practice they killed (by accident, it was said) a deer. This was unfortunate, and might have led to ill-feeling. But Snow, the CO of the RCR, had the good sense and tact to send at once a letter of apology to His Grace and along with it a quarter of the animal, which he had had his butcher prepare for cooking. After Lent (the Norfolks being Roman Catholic), the Duke and Duchess invited senior officers in the vicinity to a ball in the castle. It was a magnificent party, as formal as could be in wartime. There must have been more than two hundred guests. The hostess had invited a bevy of charming, unattached ladies as dancing partners for those of us who had not brought along our own *femmes*. I remember thinking that evening of the Duchess of Richmond's gala in Brussels for Wellington's officers on the eve of the Battle of Waterloo. There must have been a similarity, but of course we were not rushed off to do battle with Napoleon before the morning dawn!

During that summer of 1942, the 1st Brigade was moved from its coastal positions to make way for a brigade of the 2nd Division, which was earmarked for the Dieppe raid.

Much has been written about Dieppe. The plan had been approved – or acquiesced to – by the relevant senior commanders, including Montgomery. It was to take place on July 4, but was delayed because of bad weather that persisted until July 7, when the operation was cancelled. By this date the troops involved had been briefed on the location and other details of the raid. Hence the vital factor of surprise was lost. Monty, on August 10, was moved to take command of the 8th Army in Egypt, but before leaving England he made it known to General Sir Bernard Paget, commander of the forces in Britain, that, surprise having been lost, the plan should be cancelled. However, the plan was not cancelled, and on August 19 the raid took place. We know the sad result. Fifty-six officers and 830 other ranks were killed, and about 2,000 were taken prisoner from the Canadian force of 5,000.

During my time in London in 1946-48, I frequently walked through St. James's Park after lunch. One day I spotted Monty, then Chief of the

Imperial General Staff, who was also out for a stroll after lunch. We walked along together toward Whitehall and the War Office, exchanging small talk, when Monty suddenly said, "You know, Graham, I think many Canadians blame me for the heavy losses at Dieppe, and they shouldn't. When the operation was once put off, I was definite and adamant that it must not be remounted because surprise was lost."

I believe that Dieppe was a disaster because it was ill-conceived from the start. Loss of surprise may have been a factor, and failure to supply the planned naval support certainly was. As Lord Lovat, the commando commander in the Dieppe operation, wrote in his memoirs, "There was a failure to correctly assess the likely enemy response to such a raid – failure to take into account the sagacity and experience of the German Commanders, Rundstedt and Haase; and the futility of asking conventionally trained Infantry troops as opposed to specially trained Commandos – to undertake such an operation, relying on tanks and combatants within only a few hours of landing." So much for the Dieppe raid, which was the reason for the 1st Division leaving their coastal positions to enable the 2nd Division units involved in the raid to be near the beaches for rehearsals and eventual embarkation.

The lovely summer of 1942 passed with more exercises, more courses, more marriages, and a few problems, but none serious.

I had commanded my unit for almost two years, an unusually long period to hold command in time of war. But Lieutenant-Colonel Jack Sprague, who had commanded his regiment, The Queen's Own Rifles, for the same length of time, and I used to joke with each other that we must be too good to be fired but not good enough to be promoted! However, in early September, 1942, I received an invitation from General McNaughton, the Canadian Army commander, to lunch at Army HQ. My brigadier, Rod Keller, phoned me to say that he, too, had been invited and that we were both going to be told of our promotions. He asked me if I would like to ride up to Army HQ with him. And so off we went on September 8, 1942, to learn what our futures were to be.

8

FROM ENGLAND TO SICILY

The Army commander made his announcement about the promotion and move of a dozen or so senior officers at a luncheon party at Army HQ near London. Among the appointments were those of Rod Keller to be major-general and posted from the 1st Brigade to command the 3rd Division, and Howard Graham to be brigadier and posted from commander of the Hastings & Prince Edward Regiment to command the 7th Infantry Brigade. I was to succeed Harry Salmon, my old commanding officer, who had been in command of the 7th Brigade for many months and would now be a major-general commanding the 1st Canadian Division. Knowing Harry of old, I appreciated that I would be taking over a well-trained formation and would have my work cut out to maintain his high standards. Since the 7th Brigade was in the 3rd Division, I would be serving under my former brigade commander, Rod Keller, and for this I was very glad.

After the lunch, Salmon drove me to Hodgson's, the military outfitters just off Piccadilly Circus, where we both had our new rank badges sewed on tunics and cap. He then drove me back to my regiment and he continued on to his 7th Brigade. The next morning I asked Major Sutcliffe, who was to succeed me as commanding officer, to muster the regiment in a hollow square so that I could say farewell. It would be insincere to say that I did not welcome the promotion, but there was no joy in my heart as I walked forward to say a few words of appreciation for the support all ranks had given me during my two years of command. I was leaving a unit of which I had been a part and to which I had given time, loyal service, and funds for twenty years, in times of peace and war. I was leaving officers, NCOs, and men from Major Sutcliffe to private soldiers, many of whom I had known and whose parents I had known for many decades. As I thought of our days together, my voice broke and I could not go on. With a weak wave of my hand, I stood silent as they gave me three rousing cheers, and then I walked away to my car, which would take me to my new command.

On arriving at 7th Brigade HQ in Storrington, West Sussex, I was met by Salmon, who had the whole brigade assembled nearby. After a word or two of farewell, he introduced me. Having been on the receiving end of similar scenes in World War I, I took less than a minute to say I was honoured to command such a splendid formation and looked forward to their support and co-operation in maintaining the high standard set by Major-General Salmon. I then called for three cheers for their departing brigadier, and the show was over in less than five minutes. I think it was nicely done. My units in 7th Brigade were from western Canada – the Canadian Scottish from Victoria, the Regina Rifles, and the Royal Winnipeg Rifles, plus support and administrative units or detachments like Signals, Engineers, Army Service Corps, Medicals, etc. Old soldiers will appreciate the little problem of "timing" when we had a formal brigade parade calling for a march past. The two rifle regiments marched at 140 paces to the minute and sometimes went past the saluting base at the double, with rifles at the "trail." The Scottish marched at 110 to the minute, with the lovely music of the pipes and drums to keep them at this sedate pace.

I spent the next four months, September through December, 1942, doing much the same work as I had during the past three years – training, attending or conducting exercises, lecturing, inspecting, supervising inter-unit competitions in everything from sports to rifle and machine-gun shooting, cliff-scaling with alpine equipment, and overseeing combined operation and commando training in Scotland. I was fortunate to be able to spend Christmas with Leonard Pierce and his family at the home near Bognor. Leonard was a chemist of some note and a member of the Chemical Control Board. A day or two before the end of the year, I called a meeting of all officers of the brigade to be held in a large hall in Storrington. There I reviewed the activities of the past year and attempted to forecast what might happen in 1943. In doing this I made use of the words of Mr. Churchill, which went something like this: "For the past three years, we have been, as it were, scaling a cliff that stands in the way of our goal; first succeeding a bit, then falling back, then persisting, and by supreme effort and determination, clawing our way up until we are at last at the very edge where we can see the object of our endeavours – in the coming year we must go over the top and finish the job." In this forecast, I was partly right, in that a Canadian Corps would be engaged against the enemy in the Mediterranean theatre by mid-1943, but the officers of 7th Brigade, to whom I was then speaking, would have to wait still another eighteen months, until June, 1944, before they would "go over the top" and be landed on the beaches of Normandy.

The new year had scarcely dawned when I received a signal message that read: "From Lt.-Gen. McNaughton to Brig. Graham. Congratulations on your posting to command 1st Cdn. Inf. Bde. Vice Brig. Simonds with effect this date. Brig. Churchill Mann will assume command 7th Bde." This was unusual in two respects – first, that I would receive such a signal from the Army Commander instead of through the normal channels, i.e., from my Division Commander; second, that I would be moved from one brigade to another. I could not and cannot recall any similar transfer, nor have I ever learned the reason for it. I have concluded, however, that when it was decided that Guy Simonds would leave the 1st Brigade (which he had commanded only a short time), Major-General Salmon had requested that I be Simond's replacement. Though I was happy with the 7th Brigade, it can be appreciated that I was delighted to return to 1st Brigade, where I had served since mobilization, and in which were not only my own old regiment but also the two other excellent infantry units, the Royal Canadian Regiment and the 48th Highlanders, whose officers I knew so well. Major-General Salmon would be my boss as he had been when I was his second-in-command for six months in 1940 in the Hastings & Prince Edward Regiment. Hence I was returning to comrades of longstanding.

On receiving McNaughton's message, I at once contacted Brigadier Mann and we arranged a time when he would come to Storrington and I would formally hand over command of 7th Brigade to him. This was done, and off I went to Uckfield, some forty miles away, where was located the HQ of 1st Brigade. I had heard nothing from Simonds, and strangely enough I had never met him, but I had heard comments about his attitude toward and relationship with other officers, both junior and senior to him. He was said to be cool, reserved (perhaps shy), abrupt, didactic, a serious and dedicated student of military science and unquestionably an able staff officer. When I arrived at Uckfield, I was told that Brigadier Simonds had left the day before to be senior staff officer at Army HQ (or Corps HQ). I did not hear from him or meet him until a few months later, nor did General Salmon ever mention him. Having at a later date gotten to know Guy Simonds very well, I am sure it must have been gall and wormwood for him to take orders from Harry Salmon during the few months he served under him.

The first few months of 1943 slipped by as we continued the stiff training that Monty had introduced and Salmon heartily endorsed. The whole division had been put through a severe test of combined operations and commando training in Scotland and was now stationed in East Sussex around and about the city of Eastbourne. Night training was always considered important, and one week Salmon directed that we would simply reverse the clock, so to speak. Reveille would be at 6:30 p.m. instead of 6:30 a.m., breakfast at 7:00 p.m., noon break at midnight, etc. The joker was that we would carry on our program of training as it had already been planned for daytime, except for weapon-firing. I thought this was a stupid order and would lead to frustration and waste of time, which it did. My concept of night training was to confine it to the type of activity one would have to carry on when engaging the enemy, e.g., moving by foot or vehicle from one point to another, establishing and finding a start line for an attack, moving forward against an enemy position, or retreating (a retrograde movement to the rear, as the Americans called it), etc. I am sure the troops also thought much of the work we did or attempted to do during that week was nonsense and a waste of effort.

My years during World War I, when I served as a private, taught me much of importance that other officers, without that experience, missed. For example, I never subscribed to the poet's dictum that it is not within the private soldier's purview to "reason why." I always believed and acted on the assumption that every soldier should know why he is being asked or ordered to carry out a certain duty or go through a type of training. He might not agree with the reason given but, nevertheless, it should be explained to him. I believe that a week of night training, such as we experienced, tends to lower morale and induce a soldier to lose confidence in the wisdom and ability of his commander.

In April, 1943, we learned that the 1st Division was to take part in a combined operation "somewhere." Operations in North Africa were drawing to a close, and the general assumption was that the next move would be to someplace "in the soft underbelly of Europe," as Churchill put it. Whether our attack would fall on Sardinia or Sicily or Italy or Greece was known at first to very few. Our division commander was one who had been informed and he held numerous exercises on cloth models of the terrain to refresh senior officers on the principles to be followed in making an assault landing across beaches defended by an enemy. There was no hint as to where the assault would take place. Then the brigade commanders, and senior artillery, engineers, signals, and other officers of administrative services were allotted space in an office building (Norfolk House) in St. James' Square, London, which for many months had been used by the British as a central, very top-secret HQ for the planning of combined operations.

With the commanders of the 2nd and 3rd Brigades, Chris Vokes and Howard Penhale, and our brigade majors, I joined Major-General Salmon and his senior staff officers in Norfolk House, where we were briefed on the forthcoming role of the 1st Canadian Division in its assault on Sicily. We were to be a part of the 8th Army under General Montgomery. Also taking part in the assault would be the American 7th Army. Monty came to London from North Africa, and one afternoon at the War Office, with the aid of maps of Sicily, he explained in detail the plans for the landing and subsequent operations by both British and Americans involving our 1st Canadian Division. Some of the American force would come from the United States and join their comrades from North Africa on the night of the assault, and the 1st Canadian Division from England would join the British from North Africa the same night. It can be appreciated that the sailor boys had a tough task on their hands to co-ordinate the movement of convoys from North America, Great Britain, North Africa, and Malta so that all would arrive at their proper places at the proper time. I am told that some three thousand ships of all types were involved, from small freighters to great troop-carrying liners, warships, and monitors.

The 1st Canadian Division was to sail from Britain in two groups. One, known as the slow convoy, would carry vehicles, equipment, and supplies that would be needed immediately after the landing of the assault troops. The ships of this slow convoy would assemble at Liverpool, be loaded, and sail some days before the fast convoy. This fast convoy would assemble in the Clyde and adjacent lochs in Scotland and would carry the assault troops with their personal weapons. The two convoys were timed to sail so that they would meet some seven miles off the southeast coast of Sicily (known as the release position) just before the time of the assault, which was set for July 10 before daybreak. A few days after the slow and fast convoys sailed from Britain, a third would follow and carry reinforcements, field hospitals, supplies, and equipment to a base in North Africa. These would be needed to support the division in the ensuing battles and planned advance through Sicily and into Italy.

Detailed planning was necessary to get the right articles in the proper ships and in the proper order so that what was first needed would be readily available. In addition to the study of maps and air photos of the terrain we would be fighting over, and the preparation of operation orders for the assaulting units, it was essential at the same time to see that battalions and supporting units, such as Signallers, Engineers, and Army Service Corps stationed in the south of England, delivered their items, large and small, to the proper ports and in the proper ships. At Norfolk House we were fortunate in having the assistance of British officers and substaff with experience in this type of operation. The planning and the loading of the ships was without fault, as proven by subsequent events. All was going well with the planning at Norfolk House when Major-General Salmon received orders that he and his principal General Staff Officer (GSOI) and Chief Administrative Officer were to report to an RAF station in the west of England to fly to North Africa for a conference with Monty and other divisional commanders and senior officers to co-ordinate the plans for the invasion of Sicily. Off went the GOC with his officers.

Our offices in Norfolk House were never without a senior officer on duty. The day of General Salmon's departure I was on duty. At noon the phone rang. I answered it and the voice at the other end said "scramble," which meant using the top-secret phone that would "unscramble" a message. I did and I was stunned to be told by the same voice calling from the RAF station in the west of England that the plane with Salmon and others on board had crashed on takeoff for Africa a few minutes before and all had perished. What to do? Fearing the call might be a hoax, I phoned the RAF station and confirmed the news. Major Peter Bingham, an officer on our planning staff, was on duty with me and I sent him across to Canadian Military Headquarters at Trafalgar Square, only a few blocks away, to pass the news to Major-General Price Montague, the senior officer there, or if he was not in his office, to the senior officer on duty. From then on the matter was out of my hands. Naturally, I wondered who would succeed Salmon, my former CO and a friend of many years.

We were not left long in doubt. Within a few hours, Major-General Guy Simonds arrived from the 2nd Canadian Division, where he had replaced Major-General J.H. Roberts shortly after the ill-fated Dieppe operation. Like Simonds, my fellow brigadiers, Christopher (Chris) Vokes and Howard (Pen) Penhale, were Regular Army officers and all were graduates of the Royal Military College. Simonds at once called a meeting and had each of us outline our brigade plan for the landings in Sicily. Salmon had already reviewed our plans, but a new GOC, of necessity, would want to know details of the plans and either approve or amend them and put himself in the picture.

Since my 1st Brigade was to be on the right flank of the operation, I was called upon first. My plan involved the securing of a bridgehead and debouching therefrom to capture a small airfield near the coast and pursue and capture or destroy the enemy to the limit of our capacity. Simonds did not like my plan. It was "a waste of time," as he put it, to secure a bridgehead, and he gave me a stern lecture on the necessity of getting on with the job of beating up the enemy. I must confess that I was a bit chagrined at this abrupt and harsh criticism given in the presence of a large number of officers, especially since I had always been taught by Salmon and others that securing a bridgehead was a wise and proper operation to minimize the risk of being driven back into the sea by an immediate enemy counterattack. My friend Chris Vokes, who was to assault the beaches on my left, benefited from my discomfiture. He delivered his plan sans bridgehead, but strong on chasing "the bloody Hun" and/or "the Ities." Good marks for Chris. Pen had 3rd Brigade, which would follow the 1st and 2nd ashore. There was little that he could put in a plan until he saw how we were prospering. From that first meeting until several months later, I felt a twinge of resentment when the GOC referred to and called me Graham, but called the others Chris and Pen. Why this custom changed some months later will become apparent at a later point.

By early June our work at Norfolk House was finished. Detailed plans for each of my three regiments and attached services were printed and placed in sealed bags and labelled for each unit. Plaster models of the coastal area (the southeast corner of Sicily at the Pachino peninsula) were broken into sections like a jigsaw puzzle and sealed in large bags. Even battalion commanders did not know their destination. Secrecy had been of paramount importance, and books on Sardinia, the south coast of Italy, Greece, Crete, and Sicily were sometimes left visible at Norfolk House to confuse any "snoopers" who might be about. All the above paraphernalia was moved with the utmost caution to the appropriate ships lying in the Clyde, which would take us to the scene of operations.

Meantime, the troops for the fast convoy had been moved from the south of England to Scotland to be near their embarkation point and do rehearsals of landing operations on the Ayrshire coast. My brigade was centred around Irvine, in Burns's country, where the Division Commander came to inspect the brigade.

I found a field large enough to have all three regiments (Royal Canadian Regiment, 48th Highlanders, Hastings & Prince Edward Regiment) formed up in mass, and had invited Simonds, his ADC, and my three battalion commanders to lunch in the modest house I used as a mess. (Hereafter I shall refer to these units as the RCR, the 48th, and the Hastings – or Plough Jockeys – in order to save time and space, but I do emphasize that the boys from Prince Edward County contributed their full share to the exploits of their regiment.) It was a lovely June day; the troops, I thought, were magnificently turned out. It was a good show and all went well. We had a pleasant lunch and the GOC departed. Yet I cannot recall his having made any comment about the parade, nor did he make any reference to the few months he had commanded the brigade. He seemed preoccupied, almost morose. I was disappointed that he did not say a few words to the battalion commanders about the good appearance of their troops, but I suppose he had other, more important things on his mind.

Shortly after the inspection we left the Irvine area and embarked in the ships that in a week or two would take us to the Mediterranean. One day Admiral Louis Mountbatten came to the anchorage and, with General Simonds and the three brigade commanders and four or five other staff officers, visited the various ships that would form the convoy. I was much impressed by Mountbatten's friendly and outgoing personality. At midday we interrupted our visits to the troop ships in order to have lunch as guests of Commodore Teacher of the Royal Naval Reserve. His yacht, the Motor Vessel *Katharine*, was a beautiful little ship, and the dining saloon comfortably seated the eleven guests and host at a circular table. Need I say that we had a little nip of Teacher's Highland Cream Scotch Whisky before lunch? Our host was owner of that world-famous distillery.

My brigade headquarters were assigned to H.M.S. *Glengyle*, a medium-sized passenger vessel taken over by the Admiralty as a troop transport, and in it were also the assault companies of the Hastings & Prince Edward Regiment and detachments of other services. It was my good fortune, when I boarded *Glengyle*, to meet at once my senior naval officer landing, or SNOL. His name was Andrew Gray, a retired captain of the Royal Navy who had been recalled for service during the war. He was about my age, had served in World War I, and, as a naval lieutenant, had been awarded a DSO for leading a party of Royal Marines in an attack against the Turks. Captain Gray's responsibility, as SNOL, was to supervise and control the movement of the troops from the troop ship to the enemy shore. To do this he had to know the details of my plan for the assault on Sicily.

One reason for embarking the troops in the ships several days before we were to sail was to enable us to do rehearsals of landing techniques. The landing was effected by vessels called Landing Craft Assault (LCA). Each would carry twenty-five or thirty men and were slung from the parent ship's lifeboat davits (in lieu of the usual lifeboats), but because of the shape and size of the LCAs they were hung outboard and not inboard as lifeboats are. When the ship would anchor in the release position about seven miles off the enemy shore, the assault troops, with their normal weapons plus wire cutters and bangalore torpedoes to cut or blow lanes through enemy wire on the beaches, would take their prearranged places in their designated LCA. The LCAs were then to be lowered into the water and pushed off. Naval personnel, with the aid of compass bearings and outboard motors, would head for what one hoped would be the proper spot on the enemy beach. The rectangular ramp for a bow would be lowered when the craft hit the beach, and the occupants would rush out and go about their varied tasks. The LCAs would then return and be filled again by troops who would come over the side of the mother ship and scramble down into the LCAs by the use of landing nets. All these procedures were practised along the Ayrshire coast, landing on beaches similar to those that would, it was thought, be encountered in Sicily.

Two landing craft were earmarked for me, my staff, representatives of supporting arms and services, such as artillery, signals, engineers, etc., and a defence platoon. In one craft I would follow the forward assault troops to the beaches and in the other craft my brigade major and the balance of the Brigade HQ would follow the second wave of assault troops to the shore. The reason for this dividing of my HQ was to avoid the chaos that would result if the entire Brigade HQ was lost on the way in. The final rehearsal was at night. In my LCA I was approaching the shores of Scotland, south of the Clyde, when we "grounded." The naval ensign in charge of the LCA lowered the ramp and out I leapt, thinking I would be on a nice sandy beach. But alas! I was on sand, but it was covered with cold sea water up to my neck, and my staff and I were fifty yards from shore! The craft had grounded on an uncharted sandbar. Bad luck. We waded ashore, parachute lights were now being fired high into the sky, and they hung there long enough for us to see and make our way up the sandy beach to find a place to set up a temporary HQ. And where was this? In a deep sand bunker on what we later learned was the Royal Troon golf course!

First order of business was to get communications working and try to advise Division HQ, which had remained on board their ship a long way off. Our signal set was supposed to be waterproof, but some water had apparently gotten into it and we could not contact HQ. My signallers did their best and after an hour or two they were able to receive, and they said to me that General Simonds wanted to speak to me. I put the earphones on and got a hell of a blast as to why I had not reported my position, my progress, etc. I did my best to get the information to him but he could not hear me. Finally, the boys got the transmission problem cleared up and we sent the necessary reports. But by then it was mid-morning and we had re-embarked, the Navy having found an opening in the bar to get the landing craft to the beach and take us off. This was the first but it was not to be my last experience with a sandbar.

I returned to the *Glengyle* about noon, glad to get into clean, dry clothes and be free of the sand of a Troon bunker. Later that day the GOC had a conference of senior officers on board his Headquarter ship, H.M.S. *Hilary*. I should say here that the senior naval officer in the convoy was Rear Admiral Sir Philip Vian – best known as "Vian of the Cossack" because early in the war, when he was a captain commanding the destroyer H.M.S. *Cossack*, he had followed the German ship *Altmark* into a Norwegian fjord, boarded her, freed several hundred British seamen who had been made prisoners when their ships had been sunk, and then sank the German vessel. Admiral Vian sailed with Simonds and the Division HQ staff in H.M.S. *Hilary*, and of course was the senior officer to whom my SNOL reported.

Simonds, instead of asking me what my trouble had been, as I think a senior officer of wisdom and experience would do, directed a vicious

143

tirade against me because of his lack of knowledge of my whereabouts and my failure to report to Division HQ. In general, he pictured me as a most incompetent person to be commanding a brigade. I was distressed at this dressing down in the presence of many officers, both junior and senior to me, but what could I do or say? Perhaps the Navy ensign had erred in navigation when he put my craft on a sandbar, or perhaps it was an act of God, as sandbars do shift. So I sat silent. But not my SNOL. I learned then, and confirmed it later, that naval officers, though well disciplined, are prone to talk back if they feel an injustice is being done. Captain Gray rose to his feet and gave my GOC as neat a defence of my actions as any lawyer could have done. He started off by saying, "Sir, if you had asked me or Brigadier Graham what the reasons were for the events to which you have referred, you would not have found it necessary to make the criticisms you have." He then went on to explain what had happened. The GOC did not reply. I had the definite feeling that he wanted to be rid of me, a rank amateur. Imagine, a lawyer in command of one of his brigades! But it was too late now. In two or three days we would be sailing for Sicily.

In the late afternoon of June 28, H.M.S. *Glengyle* and the dozen other ships that constituted the fast convoy weighed anchor and slowly made their way out of the Clyde Estuary and fanned into three lines. The *Glengyle* was at the head of the right-hand line; H.M.S. *Hilary* was in the centre; and another ship, carrying Brigadier Vokes, headed the left line. I had the privilege of being on the bridge of the *Glengyle*, and as we headed toward the lowering sun I saw a very large camouflaged vessel speeding toward the Clyde. It was the *Queen Elizabeth* near the end of one of its many unescorted crossings from the United States, loaded to the limit with troops and equipment. Our convoy steamed westward far into the Atlantic before turning south in order to avoid possible attack from enemy shore-based aircraft. We had no close-in naval escort, but occasionally we caught a glimpse of a small ship screen of frigates on the horizon. There was radio silence, and messages were passed by lamp signals or a hoist of flags. The weather was good and the voyage was pleasant for the first week. On the fourth day at sea, the sealed bags were opened in all ships, and for the first time all ranks were told of their destination. Detailed orders and information with regard to terrain, climate, the natives, and the enemy were read and studied. The plaster models of the beach were set up, lectures were given, tropical uniforms were issued. Ointment was issued to be used to offset sunburn. In addition to daily physical exercises, troops stretched out in the sun at controlled intervals in order to acquire a proper, healthy suntan. Indeed, from morning to night, there was little time for loafing.

About seven days after sailing, Lieutenant-Colonel Sutcliffe showed me a flag the sailmaker in the *Glengyle* had made and presented to him. It was about six-feet square of gold-coloured cloth with the badge of the regiment in royal blue, very accurately done, sewn or appliquéd on the centre (the colours of the regiment are blue and gold). Captain Gray saw it and thought it would be nice if it was flown on the ship when the troops disembarked; Sutcliffe and I agreed, and Captain Gray sent a signal, by lamp, across to Admiral Vian in H.M.S. *Hilary*, telling him of the sailmaker's effort and asking for permission to fly it from "the triatic stay" (a device attached to the main mast of a ship) when the *Glengyle* was at anchor during the assault landing.

In about an hour a signal came back, not from Admiral Vian but from General Simonds, quoting from King's Regulations that no regimental colour or flag could be used by a unit until it had been approved by higher authority and sanctified by a cleric. I felt somewhat humiliated at this reply, because I knew all about the rules concerning regimental colours but had considered this gift of the sailmaker a "battle flag." My SNOL thought the same and was livid at receiving a reply to his signal from the General instead of from the Admiral and also at what he called the nit-picking objection by Simonds. But Captain Gray was not to be put down so easily. He at once sent another signal to his Admiral, saying he was not interested in the General's dictum and would like a reply from the Admiral to the message he had sent earlier. He got an immediate and typical naval response: "Keep your shirt on – you have reached age of discretion – use it." The flag was flown from the triatic stay and was later given to Lieutenant-Colonel Sutcliffe. It is highly prized by the regiment, though it is mutilated by holes from fragments of hot shrapnel and enemy mortar fire. Many times I saw it in Sicily and Italy draped over a cactus or hanging from the limb of an olive or almond tree or lying from a stone fence, with Mr. Duffy, the regimental sergeant-major, explaining its history and the history of his regiment to a new batch of reinforcements. It was never consecrated. Too bad!

After sailing southerly and well out in the Atlantic, the convoy turned east on July 6 and, in the dark of night, passed through the Strait of Gibraltar into the Mediterranean. The weather continued fair, the sea having a gentle rolling swell. I was on the bridge with Captain Errol

Turner, R.N., commander of the *Glengyle*, and Captain Gray, when far ahead we spotted a ship coming straight for us, at high speed. Within a few minutes we knew it to be a friendly destroyer. It passed quite close to us, starboard of our line of ships, then turned and came up to within a few yards starboard of H.M.S. *Hilary*. A line was shot across to the *Hilary*, a bosun's chair was rigged, and we saw someone transferred from the destroyer to the Admiral's ship. We were pondering the reason for this unexpected visitor and agreed it must be very important, indeed, perhaps a postponement of D-day, a warning of enemy air or submarine action, or a change in plan.

We were left to fret and cogitate for almost an hour, and then by lamp signal we got the answer to our questions, which went something like this: "A false beach (i.e., a sandbar) has been discovered one hundred yards from shore with ten feet of water on inside and extending across 1st Brigade front. This will prevent use of LCAs by assault companies. New plan has been made, as follows: LCTs with DUKWs will meet parent ships at release position. Assault companies will use landing nets to embark in LCTs. When LCTs are stranded on false beach, DUKWs will exit from LCTs and carry assault troops to shore. DUKWs will remain to carry follow-up companies from LCTs or LCAs across sandbar to shore." End of message. The change in plan involved landing craft tanks, which could carry several vehicles and had a square bow that could be lowered as a ramp to permit exit of vehicles, and DUKWs, or amphibious unarmoured vehicles, which can move through water by using propellers and upon reaching a beach can switch their motor power to driving rubber-tired wheels.

My SNOL and I were dumbfounded, first at the fact that the sandbar had not been discovered before this late date, and second at the decision that DUKWs would be used to circumvent the problem, for being made of wood they were vulnerable to mortar, shrapnel, and small arms fire. My assault companies had rehearsed a number of times in the LCAs, and each man knew his exact place and task. They were now to be put in a strange vehicle of different size and different shape, carrying fewer men than an LCA. In the dark the LCT must find the *Glengyle* and another ship that carried some of the assault companies and the men would have to go over the side of their parent ships by scramble nets and into the LCTs and thence into the DUKWs. For my SNOL, whose responsibility it was to get the troops safely ashore, and for me to be faced now with a completely new plan, and for my subcommanders and their men, this unpractised type of operation had all the earmarks of developing into a disaster as bad as or

146

even worse than Dieppe. What to do? Captain Gray and I soon agreed on an alternative plan that was simple, would lead to little if any confusion, and would result in little or no delay in the landings. The assault troops would go ashore exactly as had been rehearsed except that instead of going from the mother ship directly to their allotted section of beach, they would follow the 2nd Brigade assault craft and, after passing west of the sandbar, as the 2nd were doing, our craft would "side slip" to the east and once inside the bar would have a clear run to the beach. Gray sent a signal to the Admiral saying we had an alternative plan and would like to discuss it with him and the GOC. Meantime, all ships in the convoy were moving at their normal speed of about twelve knots, and the destroyer was still steaming along between the *Hilary* and the *Glengyle*. A prompt signal from the Admiral came back, "Okay, will send destroyer to fetch you." We were to have a rather hazardous adventure!

The *Glengyle* had, near her stern, large bundles of slats fastened to the hull to act as fenders. Gray and I put on rubber-soled sneakers. The destroyer came round our stern and nosed gently up to one of those fenders on the *Glengyle*, both ships continuing to move at the convoy speed. The "Med" was relatively calm, but even so there was a gentle swell of three or four feet and this, combined with the wash from the *Glengyle*'s hull, caused the bow of the destroyer to rise and fall some ten or twelve feet. Gray and I climbed out onto the fender and when the destroyer was at the top of its rise, with the swell, with its narrow deck about ten feet below us, Gray jumped. I hesitated just a fraction of a second because I did not want to collide with or land on top of him, and then I did my leap. But the destroyer started its fall just as I jumped, so I sort of followed it down to the low point of the swell, and made a good four-point landing on the narrow deck after almost twenty feet of free flight! It was not bad for a fellow a week short of his forty-fifth birthday. Helping hands hoisted me to my feet and I was shaken but not broken!

With Gray and Graham safely on board, the destroyer backed off from the *Glengyle* and over to the *Hilary*; there we did a reverse transfer. There were similar fenders on the *Hilary*. By standing on the cable that served as a handrail on the destroyer, we were able to reach up and clutch the hands of seamen lying flat on the bundle of slats, and so were hauled aboard the flagship. We had a good discussion. It was mostly between Vian and Gray. The latter emphasized the difficulties of the revised plan and insisted that there was no possible chance that my assault companies could be put ashore in accordance with the original timetable. The LCTS from Malta would find it difficult to locate my parent ships in the dark, especially with radio silence still imposed.

My principal objections were the following. First, the use of unarmoured DUKWs would invite unacceptable numbers of casualties. Second, the delay in our landings, which were bound to occur, would put at risk the assault of the 2nd Brigade on my left and the British 231 Brigade on my right. Third, there was the confusion of the men of my assault companies in being put in strange vehicles, the type of which they had never seen. Vian admitted the validity of our objections to the altered plan and agreed that our alternative proposal would be more likely to succeed and would cause much less delay and confusion and fewer casualties if the landings were seriously contested by the enemy. However, because of the short time now remaining before D-day, there could not be consultation with the planning staff in Malta, who had hatched the DUKW plans, and Vian was not prepared to ignore or disobey it in favour of our proposals. During this discussion I had no recollection of my GOC making any comment. Perhaps, rightly or wrongly, he felt that a decision as to how the troops were to be put ashore was for the Navy to make, and he had no reason to butt in. I felt somewhat lonely and dejected at his attitude and thought that the least he could have done was to commiserate with me and my problem and say, "It's too bad this had to happen, but just do the best you can, Graham." Instead there was not a word of sympathy or support. So, after being treated to a cup of coffee, we returned in our sneakers to do the jump from the *Hilary* to the destroyer (my timing was better) and, with the help of willing hands, from the destroyer to the *Glengyle*. I repaired at once to my cabin and as quickly and clearly as possible produced a message to be sent by lamp, to Lieutenant-Colonel Crowe of the RCR in the ship astern of the *Glengyle*, and for Lieutenant-Colonel Sutcliffe in my ship, telling them of the new plan and the reason for it. Crowe had been one of my company commanders and Sutcliffe my second-in-command in the Plough Jockeys. It is an interesting fact that, because of the close rapport that existed between us, there were no questions asked and no complaints from them.

We continued eastward along the North African coast in beautiful tropical weather, the fast convoy seeing no sign of enemy aircraft but learning of the presence of an enemy submarine that on July 5 or 6 sank three of our ships in the slow convoy. Later we were told by a signal from the *Hilary* that the U-boat had been sunk. Our course kept us away from

Sicily and Italy, and not until the morning of July 9 did we turn northward toward the rendezvous with craft from North Africa and North America, south of Malta. I was told that the course we had taken would tend to lead the enemy to assume that our assault was to be against Crete or Greece, because we had left Sicily and Italy some distance to the north and west. We had been favoured with fair weather from the time we left Scotland, but early on the morning of July 9, gale-force winds lashed the Mediterranean Sea into a fury of twelve-foot waves. Many of the troops were violently seasick and for a time it looked as though the landings, due to take place fifteen hours later, would have to be postponed. However, by the evening of the 9th the winds had abated, and though it needed many hours for the rough seas to subside, the invasion was to go as scheduled, with the assaulting troops to reach the beaches at 2:45 a.m. the next day, July 10.

The naval role in the invasion process was threefold. First, it had to ensure the safe and timely arrival of the assault forces at the beaches. Second, it had to cover the disembarkation by neutralizing fire from battleships on enemy shore positions. Third, it had to continue such support during the establishment of the assault forces until their own artillery support became available. Captain Gray, in consultation with me, would arrange the movement of the troops from the parent transport vessels to the beaches by use of LCAS, LCTS, and DUKWS. Captain Gray came under the orders of his Admiral, Sir Philip Vian, not under the orders of the GOC, General Guy Simonds.

The official Canadian history of the invasion of Sicily, compiled and written after interviews with participants in events and perusal of diaries, orders, messages, and other material, states that it was General Simonds who arranged for the use of DUKWS and that he had informed me by a signal that if they did not arrive in time I was to revert to the use of LCAS. How I was to get the LCAS over the bar is not stated. Were the troops, loaded with ammunition, weapons, and equipment, to debouch from the LCAS into ten feet of water and be drowned or mowed down by machine-gun fire before they could leave the water? No. I think that the army historian, in this instance Lieutenant-Colonel G.W.L. Nicholson, a careful and excellent writer, was misinformed or read a document I had never seen. Would Gray and I have gone to see Vian and Simonds and request permission to use the LCAS if we already had permission to do so if the LCTS were late in arriving, as indicated in the official history? The only order that Gray and I saw said that DUKWs were to be used for three of the assault companies and LCAs for the fourth company because its intended route to the beach would pass to the west of the bar.

Shortly before 1:00 a.m. on July 10, we were some seven miles off the southeast tip of Sicily. It was dark. The moon had not risen. Quietly the anchors were dropped. We were in the release position. From here the assault companies of the 1st Canadian Division and assault forces of the American and British forces on our left and right would make the initial attack on enemy forces in Sicily, the first stepping stone to the mainland of Europe. The troops were below deck until it was time for them to come up and climb silently into the LCAs, slung outboard from lifeboat davits. They would take the same seats they had in rehearsals; bangalore torpedoes and wire cutters were already aboard. At 1:30 a.m., the boys came topside and, as quietly as possible, stepped across from deck to craft to take their places. The craft were lowered to the water by naval personnel who manned the LCAs. They orbited or circled near the mother ship until all were down and then, in company lines, three to a company, they made for their allotted beach – hopefully, because in the dark of night it is not easy in a small craft carrying about forty men to find accurately a small area on a strange shore when the craft strikes bottom in two or three feet of water. The prow-ramp was dropped by two of the crew, and out burst the invaders; the first three or four hit the sand and raised their rifles or bren guns to protect their comrades, who raced forward with bangalores and pliers to cut paths through the wire entanglements and so allow the rest of the company to surge forward – and again, hopefully, with grenades, mortars, rifle, and bayonets to drive the enemy out and open the way for the follow-up companies.

All these activities had been practised and rehearsed and were now carried out by the 2nd Brigade and commando units on the left. Not so the 1st Brigade. Only one company of the Plough Jockeys made that sortie successfully. The other three companies (two of the RCR and one of the Hastings) waited for the LCTs with the DUKWs that were to help them in "crossing the bar."

Words fail me in trying to describe the frustration, the utter fury of Captain Gray and myself during the two hours and more between the anchoring of the ships, the dispatch of my assault companies, and my concern for Sutcliffe and Crowe and their troops. As we had anticipated, the LCTs were slow, very slow, to find us in the darkness and the great concourse of some two thousand ships. Once found, the problem was to tie up to us with great heaving seas bashing about and, once tied up, to get the men down scrambling nets with all their gear and into the bucking LCTs. I had confidence in Sutcliffe and Crowe and knew they were doing their utmost to get the assault troops away, and Gray knew and appreciated the problems the LCT Navy personnel were having. By this time, 2nd Brigade had been gone more than an hour and I knew Simonds and Vian must be beside themselves with impatience at our delay. At a little after 3:00 a.m., the Admiral could hold his patience no longer and sent a lamp signal to my SNOL: "Will your assault never start?" It did, a minute or two later, when two companies left in LCAs, having given up on ever getting into DUKWs, but the RCR company did not get away in the LCTs until nearly 4:00. The LCAs did a left hook around the bar and put their troops ashore very near their appointed places.

I had planned on taking my half of Brigade HQ ashore in an LCA as soon as my assault companies had landed. So when the Hastings & Prince Edward boys left the *Glengyle*, I asked Gray to have my craft (which was on the foredeck) lowered over the side. As a derrick swung the LCA out over the water and began to lower it, a heavy swell caught it and banged the stern against the *Glengyle*; the propeller and engine were damaged and the craft was useless. At once my second LCA was lifted and this time safely placed on the water. The Navy crew of four steadied it, while my entire Brigade HQ and I scampered down a net and took our places – very crowded we were! I sat in the stern beside the skipper, a very young naval petty officer. As we pushed off from the *Glengyle*, he asked, to my great surprise if not consternation, "Which direction do we take, sir?"

"Why," I replied, "don't tell me you don't know the bearing for 'Roger' beach?"

"No, sir, I'm afraid I don't. I was supposed to just follow the other LCA in."

It was beginning to show daylight by this time, so I said, "Do you know the bearing to the coast?"

"Oh, yes," he replied. "I can find Sicily all right, sir."

I could not help smiling at him and, patting him on the shoulder, said, "Okay, Admiral, steer for the shore and when you see a water tower, as you should in a few minutes, just steer for it and we'll all be happy. It's at the edge of an airfield near Pachino and that's where I'm bound for."

"Aye, aye, sir," he replied.

The water tower was a distinctive mark on the skyline, as shown on the plaster models of the Sicilian coast we studied in the *Glengyle*, and the airstrip about three miles from the beach was my brigade's first objective after landing. Soon we could see the tower and, shortly after, we grounded on the troublesome sandbar. DUKWS were moving back and forth to the beach and one waddled up beside my grounded LCA. Some of my staff and I clambered into it and shortly were deposited on the sands of Sicily. I had not wet my feet, yet the memory of that night has never faded. I have often said it was then that my hair turned grey!

By this time the sun was well up. The assault companies of the RCR and Hastings were well clear of the beaches, and follow-up companies were coming ashore via LCT, LCA, and DUKW. All were landed very near to their appointed places, with the exception of one company of the Hastings, which could not be found for more than two hours. Then, to my great relief, it arrived with the company commander, Major Alex Campbell, astride an ass, leading his men. This company had been landed five thousand yards west of their proper place and, in fact, at the extreme western limit of the 1st Canadian Division area. It was amazing how quickly all units collected themselves and pursued their tasks in a well co-ordinated and efficient manner. The RCR had neutralized an Italian artillery battery defending the airfield and taken several hundred prisoners. The airfield had been ploughed, as it was anticipated it would be, but we had brought with us a British engineering company with a small (D6) bulldozer. Along with a limited number of other essential vehicles, guns, armoured carriers, etc., it came ashore on the afternoon of July 10, by which time ships had moved from the release position seven miles offshore. All vehicles had been waterproofed, i.e., their engines operated in four or five feet of water.

The engineers with their bulldozer and other equipment had the airfield ready to receive aircraft from North Africa within a day or two. My reserve battalion, the 48th with Lieutenant-Colonel Ian Johnston in command, came ashore in the afternoon and moved inland from the beaches, preparatory to taking over from the RCR or Hastings next day. In spite of the sleepless night of July 9-10 and the confusion resulting from the change in plans for landing, the 1st Brigade, by last light on July 10, had captured all its objectives, had established contact with the 152 Brigade of the 51st Highland Division on our right and with the 2nd Canadian Infantry Brigade on our left, and had moved inland about four miles, having overrun a number of Italian machine-gun and mortar positions and taken several hundred prisoners – all Italians, who were disarmed and sent back under escort to the beaches. By nightfall my brigade – not least of all myself – were pretty well spent. With the 48th acting as outposts, we had no trouble dozing off for a few hours after a brew of tea and hard rations.

Before sun-up on July 11, I was out from under my greatcoat and mosquito net. I had been sleeping on the sand, and though the heat during the day was terrific, it cooled off as soon as the sun went down, so much so that some covering was desirable. Mosquito nets had been issued to all ranks, large enough to cover and protect a person from the ravenous, malaria-carrying mosquito. My jeep, my driver Corporal Baker, my signal set, and my operator had come ashore late on the 10th. So after a shave and wash and a pretty good breakfast of coffee, hard tack, and tinned bacon, warmed over the sterno, I set off with Baker and one signaller to see how the boys were up front. As was the custom, my brigade had been assigned certain roads – routes or tracks. We had just gone a short distance on one of these tracks when, suddenly, on rounding a blind turn near the beach, we were confronted by a DUKW. The track was about eight feet wide, and on each side were drains with a sharp drop of about a foot built of cut stone. The large wheel of an oxcart might drop into one of these drains and be pulled out, but the wheel of a jeep or a DUKW would probably be broken and certainly would need a jack to get it out. The DUKW, eight feet high, stopped; the jeep, three feet high, stopped. Brigadier Graham, wearing dark glasses, well covered with dust, sitting beside long-time, loyal, efficient driver Corporal Baker, lets out a roar at people in the DUKW, "What the hell are you doing on my road and going in the wrong direction?" (Traffic was to go only *away* from the beach.) Before a response could be made from the DUKW, Baker pokes his elbow in his Brig's ribs and in a low voice says, "That's General Montgomery, sir."

Whereupon the Brig gets off his ass, yanks off his dirty dark glasses, and sees before him, leaning over the front of the DUKW, with grins on their faces, not only the Army Commander, but Admiral Mountbatten, Chief of Combined Operations, and General Oliver Leese, the Corps Commander. Fluttering on the front of the DUKW was the Army Commander's pennant, which I should have noticed. "I beg your pardon, sir," I said, and left the jeep to walk up to the DUKW. "The sun was in my eyes."

"That's all right, Graham," Monty smiles, "I'm looking for Guy

Simonds." He remembered me from the days in southeast England. "Where can I find him?"

"I'll just back up, sir, and show you where his HQ are, sir." I did. We backed up a short way, where there was a culvert over the drain, and Baker could thus make way for the eminent trio. We parted with a smart salute from me and a friendly wave from Monty.

The group had come ashore a half mile east of my position. The boys liked to tell a story – apocryphal, of course, but a play on Monty's conceit – of how when the DUKW was coming ashore he saw some troops on the beach nearby, whereupon he tapped his DUKW driver on the shoulder and said, "Steer away from the troops. If they recognize me they'll wonder why I'm not walking on the water."

9

THE SICILIAN CAMPAIGN

The next three days, July 11-13, were a repetition of the 10th: marching in the heat of the hot Sicilian sun, meeting pockets of half-hearted resistance from Italian coastal units, and taking prisoners. This became something of a nuisance because when one of the three regiments took prisoners, they sent them back to my HQ under escort of an officer or senior NCO. My HQ was usually near the leading units, so their officer or NCO had but a short distance to go to deliver their POWs, but to get them back to Divisional HQ was often a long, hot, and dusty trek on foot, and I had not all that many officers available to do this chore. One day, having had a hundred or so "Ities" delivered to me, I had my brigade major select the senior Italian officer in the group and, having assured myself that all weapons had been taken from them, we sent them back to division in charge of their own officers. Unfortunately, General Leese, our Corps Commander, happened to meet such a cavalcade one day and was quite upset at my labour-saving device! Too bad. Thereafter I had to have the POWs properly (more or less) escorted.

On July 13, only three days after landing, I received word that General Montgomery would be visiting the division on the 14th and wanted to see the troops. It was not practical to have my whole brigade in one place, as I had done in Scotland when General Simonds came to inspect them. Therefore, I had each unit form up in its own area, all near the town of Giarratana, and I arranged to meet Monty at a certain map reference and lead him to the various units. It was a typical hot, dry July day. In my jeep, I met the General; he was in a camouflaged touring car or sedan with the top cut off, and with him was Colonel Trumbull Warren, a 48th Highlander officer who was one of Monty's liaison officers. I stood beside his car, saluted, said, "Good morning, sir, I will lead you to my first unit and from there go on to the second and third."

"Good morning, Graham," he responded, and with a smile continued, "but just show Trumbull on the map where the units are and we'll lead you. I don't like this Sicilian dust, you know." The foxy old fellow wasn't going to eat the Brigadier's dust! So I showed "Trum" on the map where we were to go and away we went, Baker and I choking on Monty's dust!

I have forgotten the order of visits, but the procedure at each of the three stops is the same. Battalions are drawn up in mass formation. No arms. The men are wearing shorts, shirts, and tin hats. Monty's car drives up. General salute is given. Monty, from the car, acknowledges the compliment, then says to the co, "Have the boys gather round here," and he motions around his car. The co orders, "Break ranks and rally round the Army Commander." The boys like the idea. They hurry forward. Monty says, "Sit down, boys, and take off your helmets, I want to see your good Canadian faces again. It's been quite a while since we worked together in England. I hope you are enjoying it here. They tell me it looks just like Canada." Loud laughs and ho-hos, because there isn't a blade of grass or a tree to be seen under that burning sun. Monty continues in this jocular vein. "I hope the Brigadier is getting the beer up to you every day." More laughs and ho-hos, as no one had seen a bottle of beer or of anything else since he left the *Glengyle*. Monty enjoys his jokes, too, and smiles broadly.

Then he gets down to serious matters. "I want you to know I am glad to have you in the 8th Army. You have been doing a splendid job and I know you will keep up the good work, but I want to warn you that we are going to be faced with difficult country and soon you will be running up against the Germans. So far you have met only the Italians, and they don't want to fight; but the Germans are tough, very tough opponents. But I am confident that you can master them. I will make good plans. I wouldn't be here today if I didn't make good plans. But good plans need good soldiers to carry them out. I have confidence in you. You have confidence in me, and all will be well. The best of luck to you all." Given his relationship with his troops, it is easy to see why he is a great commander. Three cheers for the Army Commander, and we drive on to the next unit. The same line is taken with the other two battalions and attached troops of my brigade, and then with a word of commendation to me, he departs.

As G.W.L. Nicholson noted in *The Canadians in Italy* (1956): "For the majority of the troops the first few days in Sicily were by no means a picnic. Had the men of the 1st Div. been less well trained, or in poorer physical condition, they would have found the assignment harder still. As it was, the process of their breaking in was swift and rigorous, effectively fitting them for the future tasks to which Montgomery had pointed."

It had been intended that my brigade would have a day and a half of rest following their strenuous four days after landing. July 14, one of the rest days, was pretty much taken up with Monty's visit. However, as often happens, events had not gone as quickly and successfully as the Corps Commander had planned and hoped they would, with the result that in the early evening of July 14, I received orders from my GOC (via a liaison officer) that my 1st Brigade would move by motor transport during the night of July 14-15 from the area of Giarratana to south of the town of Vizzini, some twelve miles to the northwest; that from there they would advance through the rather large town of Vizzini *after it had been cleared by the 51st Highland Division*; and then they would continue on a designated road, Highway 124, through the next town called Grammichele, some ten miles west of Vizzini; and finally, if possible, they would continue westward about ten miles to the town of Caltagirone.

This was an unexpected development. With no loss of time, upon the arrival of our transport vehicles about sundown, three regiments of infantry, a squadron of self-propelled 25-pounder guns, plus a detachment from the Canadian Three Rivers Tank Regiment, started the move from Giarratana to the vicinity of Vizzini, through a very tortuous mountain road. I had assigned the RCR to lead the brigade and occupy Vizzini as soon as it was cleared by 51 Division. I followed the RCR. Next in the column were the Hastings and the tanks and self-propelled guns, all of which would pass through Vizzini as soon as RCR had occupied it, while the 48th in reserve followed the Hastings ready to advance through it, when so ordered. During the movement throughout the night – fortunately it was moonlight – we frequently had to pull to the side and halt to allow the passage of ambulances and other vehicles bringing casualties back from the 51st Division and other troops engaged in clearing the town of Vizzini. In addition, our truck drivers carrying the troops had to drive slowly and with great caution, often around hairpin turns with a long drop on the down side.

Finally, an hour or two before daylight, the RCR halted. Ahead of us loomed the town of Vizzini on the edge of a cliff, or so it appeared, and the sound of gunfire was heard. This would be 51st Division troops clearing the enemy from the town. I sent one of my liaison officers to the Highland Brigade HQ only a short distance away so that he could bring me instant word when the Highlanders had finished their job and I could tell Lieutenant-Colonel Crowe with his RCR to scale the heights and go into the town. Meantime, I told Sutcliffe, just behind me, to try to get some food into his men and some rest before they started their move through the town and on to Grammichele.

It was daylight before I got word from 51st Highland Division through my liaison officer that the way was clear for the RCR to move forward and occupy the town. Crowe lost no time, in my opinion, in getting his men moving, and shortly thereafter sent me word that the British had moved out to the northeast and his RCR was established on the western outskirts of the town. I had Sutcliffe with me and saw his Hastings with attached units of the vanguard on their way through the town and on toward Grammichele ten miles to the west. It was then almost 6:00 a.m. and a blistering sun was already beating down. I was sitting by the roadside in my jeep, shortly after, unshaven and unwashed because I had been up all night, and now intended to follow the self-propelled battery that would support the Hastings if necessary, should they run into stiff enemy resistance. With me was my trusty driver Corporal Baker and a signaller in the back. We were just about to start off when a jeep drove up beside me and out stepped the GOC, Major-General Simonds. This was the first time since the landing that he had come forward to see me, and I soon knew the reason for this visit. "Why hasn't your brigade moved forward before this?" he lit into me, obviously very angry.

"They have moved forward," I replied. "The RCR are in Vizzini and the Hastings are through or going through."

"What is all the delay? My orders were for you to move through Vizzini as soon as the 51st Division had cleared it," he scolded and continued at some length.

"I assure you we haven't wasted any time, but I want to get forward now and we can continue this discussion at your 'O' Group this evening," was my carefully controlled reply, because by now I was angry also. Without further ado, I said, "Drive on, Baker," and left my GOC standing in the road.

Baker uttered two words, "That's terrible."

The signaller behind me was more voluble, saying, "Jesus, he was mad."

It is worthwhile at this point to quote from Nicholson's official history the account of the first encounter with Germans at Grammichele, July 15:

At midnight on 14-15 July a long column of motor transport carrying the 1st Brigade, led by the Royal Canadian Regiment, set off along the very secondary road twisting northward from Giarratana. Three hours later the R.C.R. deployed in Vizzini, and the Hastings and Prince Edwards moved into the lead. At 6:00 a.m. the Brigade resumed its advance, travelling now along the paved State Highway, No. 124, which followed the narrow-gauge railway line connecting Enna with the South-east. As the column rolled forward, the Infantry riding in lorries and carriers or mounted on tanks of the Three Rivers Regiment, the troops found themselves passing through a more prosperous-looking area than they had yet seen in Sicily. Much of the way the road ran through a wide upland valley, with gentle slopes rising to the high ground on either side. Fields were large, and free from trees and rock. As usual there were no houses to be seen along the route; the peasant workers followed their custom of centuries of congregating in the hilltop towns.

Such a centre was Grammichele, a community of 13,000, ten miles distance by road from Vizzini. It was built in 1683, after an earthquake, and was constructed on the unique plan of a spider web, with six roads radiating from the central piazza. The completely hexagonal town was perched on a long ridge some 250 feet above the level of the surrounding country, and thus had a commanding view of the road from the east. It was a good spot for a delaying action.

At about 9:00 the leading Canadian troops rounded a bend in the road and saw Grammichele on the sky-line two miles to the west. There was no sign of the enemy as the reconnaissance group of the Three Rivers approached the town, with the Infantry Battalion closely following. But a strong rearguard of artillery and tank detachments of the Herman Goering Division was lying in wait, and as the first vehicles reached the outermost buildings, they came under a sudden burst of fire from tank guns and anti-tank weapons of calibres reported as ranging from 20 to 88 millimetres. The fire quickly shifted to the main body of troops; a Canadian tank and three carriers were knocked out, and several vehicles destroyed.

The Infantry immediately began closing in on the town, while self-propelled guns of the Devon Yeomanry rapidly deployed from the road into the neighbouring fields to give prompt and effective support. Guided by tracer bullets fired from one of the forward carriers to indicate the enemy's positions, the Three Rivers squadrons destroyed three German tanks and a number of flak guns. In wide, sweeping movements three companies of the Hastings converged upon the town from as many directions, while the remaining company gave covering fire. As the first Canadians gained an entry within the perimetre, the enemy began to evacuate. By noon Grammichele had been cleared, and the Herman Goerings, leaving behind them a quantity of equipment and stores, were retiring westward along the highway, harassed by our artillery. This first encounter with the Germans had cost 25 Canadian casualties.

I was following the Hastings, and as soon as they came under fire I went forward to see Sutcliffe, not to interfere with his command of the operation he was faced with but to be nearby should support be needed. As the battle progressed we sat on a stone fence to view the action. Stupid we were, because the enemy easily spotted us and soon a salvo of shells fell around us. No one was hurt, but we tumbled over the stone fence and crept away to a less prominent position! Early in the afternoon I had the 48th pass through the Hasty's and pursue the enemy to the west, with their objective, the town of Galtagirone, having a population of about 30,000. The roads were mined and progress was slow, much of it by cross-country tracks so that dark had fallen when they reached the outskirts of the town and halted for the night.

Once Grammichele was cleared and the 48th were on their way, I had Baker pull off behind a cactus hedge and, with the aid of a sterno, had a wash, a shave, and a mug of tea and biscuits, and took time to think about my contretemps with my General. It was not exactly an amicable relationship. His attitude toward me, from his first appearance as Division Commander, had been cold and critical. There was never a word of commendation for me or my brigade. He was rude and harsh in the presence of seniors and juniors, embarrassing to them and humiliating to me. Perhaps he felt that I, an amateur Militia man, could not be up to the standards he required and should be treated somewhat like a schoolboy, although Chris and Pen, who were RMC graduates like Simonds, could not have been kinder or more friendly and helpful to me and had no touch of the Simonds attitude. As I rested and ruminated in the shade of a giant cactus, I decided that this morning's tongue-lashing was the last straw. I had had enough from this young gentleman, six years my junior, who obviously had no concept of how to command troops or deal with subordinates, though he might be a brilliant staff officer. It seemed quite clear that he would be glad to be rid of me, and it was only a matter of time until he felt he had sufficient justification and evidence to warrant my removal. So why wait?

That evening, at the close of his "o" Group meeting, held in the corner of a field west of Grammichele, when he gave his orders for next day's operations, the group dispersed and he started away with his intelligence officer to get in his jeep. I caught up with him and said, "Just a minute, sir! We have a discussion of this morning to be finished."

He naturally was taken back and said, "What do you mean?"

I then gave him my tale of woe – his cool, critical, and rude treatment of me on a number of occasions in the presence of senior and junior officers to their embarrassment and my humiliation; his treatment of me as though I were a schoolboy without competence as a brigade commander, which culminated in my statement, "This morning you scolded me and bawled me out and humiliated me in the presence of my driver and signaller. I feel the time has come for us to part company, and therefore I request release from my command at once. Ralph Crowe is competent to carry on as brigade commander, so I am not placing you in a difficult position."

The General listened to my tale, whether with relief at getting rid of me or with consternation at having a brigadier resign his command, I do not know. I loved my brigade, there was none better, and I was bitter and heartbroken at taking the action I did. As I recall, his only response was, "You'll get a bad report for this."

My reply was, "Maybe so, but my file will have reports from other commanders whom I have gladly served."

We parted. When I got back to my Brigade HQ vehicle, I sat down at a typewriter and typed a short note:

To Maj.-Gen. G.G. Simonds, from Brig. H.D. Graham: This confirms my request to you this evening that I be relieved of command of 1st Cdn.

Inf. Bde for the reasons I gave to you. Lt.-Col. Crowe is qualified in all respects to assume command.

Dated 15 July, 1943 – *H.D. Graham*.

I sealed this in an envelope, addressed it to Simonds, marked it Personal and Confidential, and gave it to one of my liaison officers to deliver at once to Division HQ a mile or two away. Dejected at the thought of leaving the brigade, but at the same time feeling that I had taken the proper though certainly unusual course of action, I had something to eat and crawled into my sleeping bag. So ended July 15, 1943, my forth-fifth birthday!

The next morning we had not moved because Vokes and his 2nd Brigade were passing through us to continue the divisional advance toward the city of Enna, the ancient capital of Sicily, located in the centre of the island. About 10:00 a.m., I received a message from General Simonds that I was to report to the Corps Commander General Leese at his headquarters, which was a long way back. So Baker and I packed all our kit into the jeep and off we went. I was not told to turn over command to Crowe or anyone else, and so assumed that the GOC had taken or would take care of that. To my Brigade Major, I simply said I was going back to Corps HQ and the GOC would be sending someone to take the brigade. It was late afternoon when we arrived at Corps HQ, an assembly of tents, trucks, jeeps, and caravans in a field of recently cut grain that was still lying about in stooks. I reported to General Leese and was told to "make myself comfortable" for the night and the Army Commander would be along next morning and wanted to see me. Baker was looked after by other drivers. We both had good meals, and no one seemed much interested in us, which suited me, so we bedded down, he in his blanket, I in my bedroll against a stook of oats. The next morning, July 17, shortly after 9:00 a.m., I saw Monty arrive, and for a half hour or so he was with the Corps Commander in his caravan. I was sitting a little way off in the shade of a tent when General Leese came over and simply said, "The Army Commander will see you now, Graham."

I went across to the caravan, saluted, and Monty said, "Good morning, Graham, sit down."

"Thank you, sir," I said, and sat.

"Now, what's this trouble between you and Guy Simonds?" asked the General. I gave him a concise history of our relationship as being cold, critical, harsh, and unfriendly on his part, and he seemed to have little or no understanding or appreciation of the problems to be faced by a commander in the field. My brigade was excellent, there was none better, we had accomplished everything asked of us. I then told him of the tongue-lashing Simonds had given me in front of my driver and signaller after an all-night, difficult, but successful move over the mountain roads to Vizzini, and I wound up my bleat by saying, "I simply cannot continue to command a brigade under such unfriendly and humiliating circumstances." This account only took three or four minutes.

Monty did not interrupt and when I finished, he said, "Have you ever refused to obey an order?"

"Never," I replied.

"Well, Graham," he continued, "you will go back with the Corps Commander this morning and resume command of your brigade."

"But this will be very difficult, sir," I rejoined. "I spoke very frankly to the GOC, and although I was respectful, I was quite critical."

"Don't worry about that. There will be no problem. Both of you are to blame. Just get on with the battle." These were his final words as he rose, shook hands, and gave me a little pat on the shoulder. A little later he left, after another short talk with the Corps Commander.

About 11:00 a.m., General Leese, his senior general staff officer, Brigadier-General Walsh (who had been at staff college with Chris Vokes), and a driver – who sat with me in the back seat, while Leese drove the sedan with the top cut off – left Corps HQ to visit the 1st Canadian Division. Baker followed in my jeep. About 1:00 p.m., we stopped where the road topped a mountain and had a sandwich lunch. Not a word was said about the Graham-Simonds fracas. To this day I do not know what Leese had said to Monty, but it was just possible that he supported my position and felt there was justification for the drastic action I had taken.

About mid-afternoon, we reached Division HQ. Leese told me to wait while he spoke to the GOC. I sat on the edge of a dry creekbed with my feet dangling over the side for a good half hour. Leese and Simonds were walking back and forth under some olive trees nearby, and Leese seemed to be doing most of the talking. He then departed without a word to me, and Simonds repaired to his caravan. A few minutes later he came out, walked across, and sat down beside me. "Howard," he said. For the first time it was not Graham. "Let's just forget about this whole business; you go back to your brigade and I'll tear up this note." He reached into the breast pocket of his shirt, took out my note of resignation, tore it up, and dropped it in the creekbed. And in truth, that was the end of it. Never again, directly or indirectly, was it mentioned by either of us. Guy Simonds's attitude toward me did a complete reversal. He was considerate, he treated me as he treated his other brigade commanders, and though his words of commendation may have been few, because by nature he was not an outgoing person, his criticisms, if any, were constructive and not abusive.*

Since writing the above passage, I have read the second volume of Nigel Hamilton's *Monty: Master of the Battlefield*, and in it he refers to the above incident and quotes a note Monty wrote to Lieutenant-General Leese, our Corps Commander, saying, "Graham is a splendid fellow and much loved by his Brigade – and Simonds must learn how to treat his subordinates." Hamilton also says I was "sacked" by Simonds, but that is not correct. I *resigned* my command. I have also learned that Simonds, on my departure, promoted his principal staff officer, Lieutenant-Colonel Kitching, to command my brigade, but upon my unexpected return Kitching was recalled or stopped en route to the 1st Brigade HQ and returned to his former appointment!

And so after one night away, I returned home to my 1st Brigade, which had enjoyed a day of rest, as the 2nd Brigade went through Caltagirone and continued the advance northwestward some twenty-five miles to the town of Piazza Armerina, about fifteen miles from Enna. Here, having penetrated almost to the centre of the rough and mountainous area of Sicily, the 1st Canadian Division changed direction to move eastward toward Mount Etna and the east coast at Catania, some seventy miles south of Messina.

An outline of events from July 17 to August 7 should be sufficient to indicate my brigade's involvement during these three weeks of almost continuous movement and battle in stifling heat, over difficult terrain, against an experienced and resourceful enemy. Success was achieved, but at the cost of a heavy toll of casualties in the 1st Canadian Division,

---

* Colonel Charles Stacey, one-time Chief Army historian, when I told him I was writing my memoirs, said, "Good, and I hope you will tell exactly what occurred between you and Guy Simonds. We have all heard about Monty firing people, but you are the only one I ever heard of being 'unfired.' " I was not fired; perhaps I would have been if I had not jumped the gun. I have often wondered what message Monty sent to Simonds by word of the Corps Commander. I am sure it referred to the treatment of subordinates.

particularly in the 1st and 2nd Brigades. As we changed direction from northwest to east, we moved on an axis through the towns of Valguarnera, Leonforte, Assoro, Nissoria, Agira, Regalbuto, and Adrano at the base of smouldering Mount Etna.

I had not long to wait for action. On the evening of July 17, I was ordered to pass my brigade through Penhale's 3rd Brigade, which had been held up by strong enemy positions some distance short of the town of Valguarnera. From information I gleaned from Penhale, I deduced that an effort to drive the enemy from their positions by a direct approach would be costly and doubtful of success. Hence, though the country was extremely rough, I decided on a two-prong attack, sending the RCR by direct road route toward the town, while the Hastings would do a right hook across country to high ground that overlooked the town and thus take the enemy from his flank. As I anticipated, the RCR met extremely heavy opposition on their direct approach to the town but inflicted severe losses on the enemy. Their second-in-command, Major William Pope, a brave and brilliant soldier and a nephew of Lieutenant-General Maurice Pope, in leading a sortie against enemy tanks and half-tracked vehicles, was killed and the regiment suffered heavy casualties. The Hastings, having made the difficult flanking move during the night, had reached positions from which they could look down upon the enemy. They inflicted heavy losses and before the day was out forced their withdrawal. I know this because early on the 18th, with Baker and one of my liaison officers carrying sten guns, I followed the route the Hastings had taken by using goat tracks and dry creekbeds and by clambering a series of terraces, each about four feet high, we reached the Hastings position and witnessed the destruction of enemy tanks and half-tracks by small arms fire, light mortars, and the use of a recently developed weapon – the PIAT, the Projector Infantry Anti-Tank.

Seeing the casualties in men and equipment suffered by the enemy, I ordered the 48th Highlanders to move across country, as the Hastings had done, and descend on the town from the hills to the south. This they did, after clearing a number of stubborn enemy positions. Their leading company alone claimed thirty-five Germans killed and a score of prisoners. The regiment, moving down from the high ground to the right of the Hastings, entered Valguarnera in the dead of night and found it clear of the enemy. Here I quote from the official history of the Italian campaign:

The fighting on that Sunday had been the most extensive in which the division had participated. There were 145 Canadian casualties, 40 of them fatal. Against this must be set the figures of 250 Germans and 30 Italians captured, and claims of from 180 to 240 Germans killed or wounded. Two days later Kesselring's daily report to Berlin carried a measure of unconscious tribute to the 1st Bde. "Near Valguarnera troops trained for fighting in the mountains have been mentioned. They are called Mountain Boys and probably belong to the 1st Cdn. Division."

There was little time for rest in these days. The American 7th Army under General George Patton was on our left, and the 78th British Division was on our right. We had to keep pace with them, although from all that I have been able to learn, the 1st Canadian Division had the most difficult ground to operate over, and the toughest, most experienced enemy troops to overcome. Having cleared the enemy from Valguarnera by the night of July 18, the 1st and 2nd Brigades were ordered on the evening of the 19th to capture the towns of Assoro and Leonforte respectively, some ten or twelve miles to the northeast. The 3rd Brigade was loaned to and placed under command of the 78th British Division.

It was about midnight of July 19 that Vokes and I got our brigades moving. About two miles from our starting point, the 1st Brigade moved by a narrow, fairly straight road in an east and north direction toward Assoro, while the 2nd Brigade continued their move on our left toward the town of Leonforte. The 1st Brigade's road followed a narrow-gauge railway for about four miles and then turned north. By this time it was daylight. I was following, as usual, my lead battalion, the Hastings, and following me were the 48th and RCR. As soon as we turned north I saw clearly for the first time the 1st Brigade's objective. It was a formidable sight. The ruins of an ancient castle stood high on a cliff, and to the left of this the road that should lead us to Assoro (not in sight because it was beyond the crest of the cliff) rose in a series of hairpin turns up the side of this escarpment. I had known from the map that my objective was on high ground, and that it would be difficult to reach it. Furthermore, I assumed that the road would likely be mined and blown up in parts. It certainly would be covered by fire from machine guns and mortars. And so it proved to be.

The Hastings' advance guard of the carrier platoon had gotten around only two or three of the sharp bends in the road when they came under heavy fire and were lucky to be able to spin around on their tracks and get

166

back to cover. Sutcliffe deployed his regiment to the right of the road we had been travelling over and I pulled off to the left. The remainder of the brigade deployed and waited developments. Because of the hills and ravines, they could find shelter from the fire of enemy guns, which were located behind the edge of the cliff that faced us. My signal truck and office truck drew up beside my jeep. It was then about noon on the 20th. Sutcliffe had located his battalion HQ in a small farm cottage in a grove of trees. The place reminded me of an oasis, as it was the only touch of green in a great expanse of parched stubble. I walked the few yards to his HQ to discuss with him a plan of some sort that would enable his regiment to scale the formidable 1,000-foot feature confronting us. To hope for success by moving up the tortuous road was out of the question. The regiment would be slaughtered. Standing in a hastily dug weapon pit under a tree, my friend of many years, who had been my second-in-command for the two years when I had commanded the Plough Jockeys, scrutinized with field glasses the face of the escarpment. At the highest point stood the stark ruins of an ancient Norman castle, and to the left of this, clear against the sky, was a row of cypress trees that I knew marked a burying ground. The southeast face of the cliff was almost sheer, but there was a considerable amount of scrub, and we could discern what appeared to be goat tracks in some places. We were some two or three miles from the base of our objective. With us and handling a map of the area was Sutcliffe's intelligence officer, Lieutenant Battle Cockin, a graduate of Cambridge University who had some experience in mountain climbing in the Andes. "Bat," we called him, and a better IO one could never find.

As we surveyed the scene, as the poets would say, the enemy from gun positions well beyond the escarpment were not idle. They kept up a desultory shelling of what the boys later called "Death Valley," where members of my 1st Brigade were lying doggo and thus avoiding any serious casualties. It was obvious to us that if one of my units did manage to scale the heights, they would quickly be blown off again, unless the enemy guns could be silenced. So it was that after a half hour of "appreciating the situation," to use an army term, Sutcliffe agreed that, as he had done at Valguarnera, he could try a right hook that might, with the aid of goat paths or otherwise, get to the top of the escarpment and establish his unit *above* the town of Assoro. We agreed that his assault companies must move with the least possible weight confined to weapons, ammunition, and full water bottles. The outline plan having been made and agreed to as feasible, if the element of surprise could be maintained until the crest had been reached, I said to my old friend, "Now, Bruce, I'm going to skip back to division and see the CRA [Brigadier Bruce Matthews, commander of the Royal Artillery] about plastering these enemy guns that are dropping shells around here, and will blow you off the top unless we can neutralize them."

"Right, sir, I'll get busy on the details and get everyone well fed so we can move as soon as it's dark," he replied.

Off I went in my jeep with Baker and my signaller in the back seat as usual. It was a half-hour drive back to Division HQ. We were almost there when I heard the signaller say into his microphone, "Say again?" He had received a message by headphones and wanted it repeated. It was repeated. Then he took off his phones and said, "Sir, Sunray 56 and his 10 have been killed by shell fire." Sunray 56 meant CO Hastings. Baker slowed down, thinking I might want to rush back to the place I had left twenty minutes ago. I wanted to, but even in a short campaign, as this had been, one becomes used to the possibility of death overtaking oneself or one's friends. So I motioned to him to drive on, and in a few minutes I arrived at Division HQ, gave the GOC and Brigadier Matthews an update on my positions and plans and my tragedy in losing a loyal, efficient, brave friend, and battalion commander. I emphasized that my reason for coming back to Division HQ was to make it clear that the operation against Assoro could not succeed unless a counter-battery program was developed and in practice by the next morning. Communication had been disrupted, and because of this, and the difficulty in locating gun positions from which shells could be fired with a "crest clearance" that would permit hits on enemy guns, our artillery had not been able to give the forward troops the support they would have liked. Furthermore, the enemy gun positions could not be pinpointed as they lay behind the ridge on which the ruined castle stood.

I did not linger at Division HQ but hastened back at top speed to the small farmstead where I had left Sutcliffe and Cockin an hour before. There I found Lord Tweedsmuir,* Sutcliffe's second-in-command, and

---

* I must explain the presence of a British peer with a regiment from eastern Ontario. The second-in-command of the Hastings had been Major O'Connor Fenton from Peterborough, Ontario, but just a few days before embarking in the Clyde, while riding a motorcycle on a narrow Ayrshire road, Fenton collided

the company commanders. To the assembled group I explained the plan that Sutcliffe, Cockin, and I had agreed upon, as they had been killed not long after I had left them and before Sutcliffe had a chance to discuss the plan with his officers. I confirmed Tweedsmuir as commanding officer, and asked if he was prepared to make the attempt and, indeed, thought it possible. I felt it would be unfair to this new commanding officer and to the men under his command to give a direct order for the attack if he felt it could not succeed and accepted the task with such reservation. I am glad to say he not only agreed but seemed enthusiastic. I left him to organize his force and to do the necessary detailed planning. I walked the short way across the road to my truck and advanced to the Brigade HQ vehicles. None had been damaged, but at the rear of my office truck was a shell crater about six feet deep and fifteen feet across! I was glad I had not been home!

At dusk on July 20 we buried Bruce and Bat, wrapped in blankets, in the corner of a field beside the road. Temporary wooden crosses made by the regimental pioneers would be planted at their head and, later, as was the custom, they would be lifted and moved to a permanent resting place, the Canadian military cemetery near Agira. Shortly thereafter, I saw Tweedsmuir and his men start to move silently in the dark on their cross-country, three-mile trek to the base of the thousand-foot-steep escarpment they must scale before dawn.

The story of that escalade has been told many times. As Sutcliffe, Cockin, and I had hoped, and as Tweedsmuir later agreed, the enemy had assumed that an ascent of the precipice, if not impossible, was certainly unlikely to be attempted by any sane person. In some places the face was

---

with a truck and was seriously injured. Because both Sutcliffe and I knew there were officers in the regiment capable of taking his place, we were not a little annoyed to be told that his position as second-in-command would be filled by Major John Buchan, the second Baron Tweedsmuir, son of the late Governor General of Canada. So far as I knew, John Tweedsmuir, as he liked to be called, had never commanded anything, had never seen the Plough Jockeys, nor had I or anyone else in the brigade ever set eyes on him. However, our fears as to his competence were soon allayed, and at once "Long John," as the troops referred to him, became a highly respected member of the Hastings. Tweedsmuir was an outdoors man who had travelled widely and, of much importance, had scaled heights in his native Scotland and in western Canada.

terraced, as we had experienced at Valguarnera, for the growing of grape vines, and the men, even though carrying only light weapons and ammunition, were almost at the limit of their endurance before reaching the crest. Part way up they encountered a deep and steep-sided ravine or fissure that ran across the face and which we had not been able to discern when studying the ground with our binoculars. Getting into and out of this unexpected obstacle was in itself a test of stamina and determination. At one point, I am told, the soldiers seemed too exhausted to continue, but the urging and example of their gallant commander and his equally determined officers and the fear of being caught in daylight like helpless steeplejacks brought forth that last bit of effort, which can so often be mustered up when one's life is in the balance, and brought them to the crest.

So it was that as the early light of day spread across the land, they found themselves on fairly level grassy slopes beside the ancient square keep of the castle ruins that we had seen from Death Valley, far below. The encounter with the few Germans who were manning an observation post was quick and deadly, but not without rifle and bren-gun fire. This was the group, no doubt, who had directed the artillery fire that had killed Sutcliffe and Cockin. A day later, when I managed to get to the crest by use of the road, I could clearly see below me the little copse and building where we had stood in surveying the cliff and discussing our plan. I could not help being reminded of Wolfe and his troops clambering up the steep hill on the shores of the St. Lawrence to reach the Plains of Abraham and surprise Montcalm and his gallant French garrison.

From this new position, Tweedsmuir and his men looked down on the roofs and streets of Assoro and commanded a clear view of the road running north from Assoro to join the road from Leonforte. From this point there was also a clear view of much of the tortuous road up which I would later move from Death Valley. As we had anticipated, little time was lost by the enemy in launching counterattacks, supported by their guns located north of the Assoro-Leonforte road junction. By good fortune, in destroying the observation post at the castle, the regiment acquired a splendid, high-powered pair of scissor-type binoculars. Major Burt Kennedy, former "D" Company commander, and now second-in-command of the Hastings, had, in his Militia days in Owen Sound, Ontario, spent some time training with a Militia field artillery regiment. Now he was able to act as a forward observation officer and, with the use of the binoculars, he was able to send back to our Division Artillery the exact locations of enemy artillery. As a result, the Germans suffered heavy losses from our counter-battery fire. This undoubtedly was an important factor in forcing the enemy to release his hold on both Leonforte and Assoro.

In the afternoon of July 21, I saw Lieutenant-Colonel Ian Johnston, commanding the 48th Highlanders, and put him in the picture as to the Hastings position. It was agreed that his regiment would attempt that night an ascent to Assoro by or more or less on the axis of the road that wound up the south side of the escarpment. Upon passing through Assoro, he would move to a position at the junction of the road running from Leonforte to Nissoria. My appreciation was that the Hastings, now certainly the main concern of the enemy, would draw his fire and his strength away from the axis of the road, and so permit the 48th Highlanders to be the left claw of a pincer that would either destroy the enemy or cause him to retire. This would open the way for the 2nd Brigade on my left, which was still struggling against the German defence of Leonforte. Ian Johnston was a lawyer and Militia soldier, as I had been in the 1930s. He was a most loyal and able commander and later he led a brigade with great distinction. He instilled confidence in his men by his calm, assured demeanour. How lucky could a brigade commander be with such regimental commanders as Johnston, Sutcliffe, Crowe, and Tweedsmuir.

To accomplish the difficult task before him, Johnston lost no time in doing what reconnaissance he could, most of it by binoculars, and agreeing that while some of the ascent might be by road, it was more likely that climbing by goat tracks or up the terraced hillside to debouch and attack Assoro from the west, as the Hastings were doing from the higher pinnacle on the east, would lead to success. The actual movement started after dark, as had been the case with the Hastings the night before. By dawn the 48th were at the gates of Assoro, and by midday, with little loss to themselves but inflicting many casualties on the enemy, in concert with the Hastings, they forced the enemy to withdraw along the main road toward Nissoria. This enabled the 2nd Brigade to clear Leonforte.

After making the ascent on the night of July 20-21 and beating off repeated counterattacks and suffering from artillery shelling, the Hastings were at the point of desperation and exhaustion by nightfall of the 21st. They were almost out of ammunition, and they had not slept or eaten for twenty-four hours. Although they caused heavy casualties on enemy personnel and vehicles and had been instrumental in knocking out several enemy guns, they had not the support or physical strength to assault the enemy below them, who were sheltered in the old stone buildings at Assoro. As the sun was setting, Major Bill Stockloser, a Hastings company commander, and Angus Duffy, the regimental sergeant-major, approached me. They had come from the Hastings position at the castle to my truck to give me a report on the situation and to lead a party with food and ammunition up the way they had gone the night before to the relief of the regiment. Lieutenant-Colonel Crowe of the RCR was with me at the time and at once volunteered to take a party of men from his unit on this onerous and perilous journey. In short order, he had a hundred volunteers to do the job, and at dusk off they went, laden with the much-needed supplies. Stockloser and Duffy led the way, and by morning light the party had delivered the goods and were back to their regiment. It was gratifying and heart-warming to have this done on a voluntary basis, and not have to issue orders, as I would normally have had to do. Here is how the Army's historian described it in *From Paschino to Ortona*: "On the 22nd, the Highlanders made contact with the Hastings and the latter unit's position thereafter was much easier. The enemy withdrew from the town of Assoro about midday, and one of the most dramatic episodes of the [Sicilian] campaign came to a victorious conclusion."

I was able to get up to Assoro shortly after noon on the 22nd and saw the bodies of several German soldiers still lying in the main street of the little town, their comrades not having had the time or opportunity to bury them or take them with them in their hurried departure. I also had a few words with Major Kennedy, who had found a quite comfortable bed in a deserted house and was recovering from a state of complete exhaustion. After almost seventy-two sleepless hours, the whole regiment was ready for a day or two of rest. But it was not to be. Our official history reads:

The seizure of the Assoro pinnacle by the Hastings & Prince Edward Regiment was as significant to the Canadian Division's advance as it was dramatic, for it upset the whole German plan of defence on that front, and thus hastened the fall of Leonforte. Assoro and Leonforte were two phases of the same battle: for the enemy had to hold the whole ridge or withdraw from it altogether. We have seen how his grasp in the eastern end was pried loose by the 1st Brigade.

Canadians in Italy, the official history by Lieutenant-Colonel G.W.L. Nicholson, also includes the following praise from an unexpected direction: "Some six weeks later in preparing his 'experience report' of the campaign in Italy the Commander of the 15th Panzer Grenadier Division paid this tribute to our troops. 'In field craft, superior to our troops. Very mobile at night, surprise break-ins, clever infiltration at night with small groups between our strongpoints.'"

Later in the day, the road up the escarpment having been repaired and swept for mines, I took Major-General Simonds to the top in my jeep, and as we rounded one of the hairpin turns, we beheld a character in very short shorts, no shirt, a rifle slung over his back, crowned with a black-silk top hat. This creature was on traffic control, and I recognized him as one of my former Plough Jockeys from North Hastings! Being well-disciplined, he raised his silk topper and bowed from the waist. I could not help smiling at his spirit and attitude, and gave him the v for Victory sign. But the GOC was not amused, and a few days later an order was issued that troops were to be properly dressed at all times!

On the afternoon of July 22, Major-General Simonds ordered the 1st Brigade to advance along Highway 117 and capture Agira, an ancient town some twelve miles to the east, perched high upon its mountain cone overlooking the valleys of the then dried-up Salso and Dittano rivers. But the way was not clear for a run at Agira. The enemy had withdrawn from Assoro only a few hours before. About five miles from our starting point was the small town of Nissoria. Because it lay in fairly flat ground between two ridges running from north to south, the GOC and his planners evidently thought it would not present a serious obstacle to my brigade's advance.

I was not so optimistic, having had experience with the German tactics of delaying the Allied advance by stationing a minimum force of mobile troops and guns on half-tracked vehicles at innumerable strategic points at four- or five-mile intervals. By so doing, they forced us to deploy a force ten times their own numbers. I had grave doubts about the feasibility of a plan that assumed that I could advance ten or twelve miles without serious opposition. However, "orders is orders," as the disciplined soldier says, so we prepared for the advance to Agira with the RCR in the lead, supported by a squadron of the Three Rivers Tank Regiment and the 2nd Field Regiment. For reasons not connected with my brigade, this hoped-for dash to Agira was delayed, and on the morning of July 24 the GOC held another conference and gave new and more detailed orders for the capture of Agira by nightfall!

The advance of my brigade was to be closely co-ordinated with a timed program of artillery concentrations on successive targets along the highway. A creeping smoke barrage would provide a screen two thousand yards long, one thousand yards ahead of the forward troops. In front of this curtain, the Desert Air Force would bomb and strafe pre-arranged, anticipated targets along the road, and six squadrons of medium bombers would attack Agira and vicinity. After taking Agira, my brigade would exploit beyond the town. When I heard this plan, I said to myself, "My God! The man must be crazy." In all my military teaching, limited as it was perhaps, this sort of wishful thinking seemed to me not only ridiculous but almost certain to be perilous and wasteful in men and material.

The "timed program of artillery concentrations on successive targets" was based on two assumptions: that the infantry, when the concentration of fire ceased, would be in a position to at once assault the enemy while he was still dazed, if not dead; that the targets were necessarily where the enemy was positioned and not merely where the planners and the GOC thought they might be. If the attacking infantry were delayed – and they were almost sure to be for some reason, e.g., enemy resistance, mined or cratered roads, or roadblocks – all subsequent pre-arranged concentrations of artillery fire would be wasted because the hoped-for infantry follow-up had been delayed. I had been taught that artillery concentrations should be brought down *on call* when needed on specified targets. To shoot around the country in the hope of hitting something, and to time this shooting on the assumption that the people you want to support will be within one hundred yards of the shells when they hit the ground, is nonsense. The part of the plan referring to "a creeping smoke barrage" was extremely dangerous because of the possibility of a strong wind or a change in the wind. (In fact, because of a strong breeze, the smoke screen did prove a failure.)

In the official history, it was stated that one of the divisional intelligence staff remarked later that the infantry had a tendency to advance too late after the artillery shelling had ceased. This failure to follow the fire support closely enough was afterwards cited by Major-General Simonds as "the fundamental cause of the reverses suffered by units of the 1st Brigade in their attacks of Nissoria." I dispute that categorically, and can only assume that the General was confused between a timed barrage program, where the troops move from a pre-arranged start line and follow the barrage, and this pre-arranged concentration program. I still resent the effort to justify a bad plan by putting blame for its failure on the backs of the troops. (Incidentally, the Air Force bombing took place along the road but, so far as I could learn, no damage was done. The artillery concentrations in general were confined or near to the road and not to the north of the road, where the strongest enemy positions, which caused us the most trouble, were located.)

Whether I liked the GOC's plan or not, an honest attempt had to be made to carry it out. It was the RCR's turn to lead the advance, and shortly after noon of the 24th, supported by a squadron of tanks of the Three Rivers Regiment, Colonel Crowe and his men set forth, followed by myself in a tank – my first and last attempt to command from such a vehicle. I had warned Tweedsmuir that his Plough Jockeys would pass through the RCR if and when necessary, and had told Colonel Ian Johnston of the 48th Highlanders to keep well up on the Hastings, as my brigade reserve. The road was under observation by the enemy from high ground to the north, and we weathered a bit of mortar fire but nothing to cause much delay. By about 4:00 p.m., the leading troops were in the village of Nissoria and had met practically no resistance. But as soon as they emerged from the eastern edge of the closely built stone dwellings, all hell broke loose. The Germans were on the slopes overlooking the town and had been waiting for the RCR to come into the open. Mortars, machine guns, and artillery swept the exits from the town. When our tanks emerged, they were met with anti-tank fire from concealed positions. From my view of the ground after the slopes were cleared a day later, it was obvious that our pre-arranged artillery concentration had, indeed, missed the main enemy positions. In spite of the might of enemy fire, Colonel Crowe and his Royal Canadian Regiment did a magnificent job. In making use of cover and by moving to the south of the road, two companies were able to outflank the enemy and got well to his rear. Unfortunately, as so often happened in that rugged country, radio communication with Battalion HQ failed, and Crowe lost contact with these companies. In an effort to locate them, he and his signaller got separated from the rest of his unit by a short distance, and both of them were hit by enemy machine-gun fire and mortally wounded. Command of the RCR then fell to Major Thomas Powers.

While this bitter battle was raging, I was in my tank in the town, and on hearing that Crowe was killed, I told Powers to do his best to regain contact with the two companies that Crowe had been trying to contact and consolidate for the night on the north edge of Nissoria. By this time it was dark and I made my way back to my forward Brigade HQ on the slope west of the town. It seemed to me that if the enemy was to be dislodged from his prepared position east of Nissoria, we must launch a fresh attack the next morning with the support of the full Division Artillery and all available tanks. The RCR had suffered about forty-five casualties, fifteen killed and thirty wounded, and the new attack should be carried out by the Hastings and 48th, with the RCR brought back to reserve. An alternative would be to have the RCR hold where they were and the next day do a flanking attack as we had done at Valguarnera and Assoro.

Certainly I wanted to discuss the situation with Major-General Simonds. I got Lieutenant-Colonel Kitching, the senior general staff officer at Divisional HQ, on the wireless and said that we had run into stiff resistance at the eastern outskirts of Nissoria and would like the GOC's agreement to my standing fast for the night and making a further attempt with additional artillery support the next morning. However, Kitching said the GOC was asleep and he did not want to waken him, so I should follow the plan and push on. Perhaps the enemy would retire. I felt that the RCR had exhausted themselves, and if any pushing was to be done that night it must be by the Hastings. To ask a battalion to move to and attack a strange position in the dark of night was against my better judgement, but orders are orders and there was just the possibility that the enemy had made a withdrawal or would do so if confronted by a fresh attack. He had done this on previous occasions. Meanwhile, the two RCR companies that had lost contact with the balance of their unit, being in danger of assault and capture by the Germans, prudently found their way back to their unit.

I met Tweedsmuir on the road some distance west of Nissoria and walked beside him at the head of his regiment, explaining in detail what the situation was and warning him that I did not want him to suffer heavy casualties, and that if he still met stiff opposition he was to stand fast until daybreak when we could then decide on a new plan. I went with him right into the town and then walked back to my advanced HQ near the 48th, and for the rest of the night sat in a cave beside Ian Johnston – and no doubt dozed a bit!

By dawn, I learned the Plough Jockeys had been able to make little headway. Tweedsmuir had been wounded by a mortar bomb and, before being taken back to a forward dressing station, turned command over to Major Kennedy. Long John was eventually evacuated to a hospital and did not rejoin the regiment until well into Italy. Other casualties totalled five officers and sixty-five other ranks. To me it was quite evident, as it had been earlier, that a full-scale, carefully co-ordinated attack by artillery, tanks, and infantry was necessary to clear the way to Agira. But the GOC insisted that one more try should be made by the 48th, which I had been holding in reserve for exploitation if the enemy withdrew. So it was that the 48th were committed, but after a most vigorous and gallant attempt, they, too, were repulsed after suffering some sixty casualties. I had now committed all three battalions, and though the enemy had undoubtedly suffered severely, they still held formidable positions on the slopes east of Nissoria.

On the evening of the 26th, the 2nd Brigade, preceded by the concentrated fire of four field regiments, one medium regiment, and a group of 4.2 mortars, launched the full-scale co-ordinated attack that should have been done, as I had urged, as soon as the RCR identified the position and strength of the enemy on the slopes, both forward and reverse, immediately east of Nissoria. The 2nd Brigade's attack was successful and exploited to and around the pinnacle of Agira. These particular operations were not only of tremendous importance to me personally but they were key battles in the task of relieving the British forces, who had been attempting to drive north along the eastern coast of Sicily to Catania. With the three Canadian brigades alternating in the advance, the 3rd Brigade having rejoined us, we continued through Agira and Regalbuto, which had been demolished by our air attacks, to the outskirts of Aderno by August 6, and here our campaign in Sicily ended. We were then placed in Army Reserve. It must not be assumed, however, that these last few miles were not contested by the Germans. They continued to offer resistance and force deployment and delays and inflict some casualties on us, but it is fair to say that after the fall of Assoro-Leonforte and Nissoria, they realized that their fate was sealed and defeat was certain.

In less than four weeks of operations, the Canadian casualties had been

173 officers and 2,261 other ranks, and of these 38 officers and 447 other ranks were killed or died of wounds. My brigade had suffered severely, particularly in the loss of senior officers. For three or four days we enjoyed rest and relaxation in the orange groves that prosper in the shadow of Mount Etna, whose black chunks of recently spewed lava were but a mile away. It was a pleasant and welcome change from the hills and cliffs and valleys and dried riverbeds we had struggled and fought through for the past month. We were now ordered to move to an area some fifty miles south between the villages of Militello and Scordia to rest and refit for the invasion of Italy in three weeks' time.

There is no gainsaying the extreme difficulties posed by the terrain and the climate. Sicily was certainly an ideal country for defensive tactics. The enemy could and did compel us to deploy a much larger force than he had for us to dislodge him. We Canadians found the heat, the dust, and the dirt of roads or tracks at times almost unbearable. The enemy, acclimatized from service in the deserts of North Africa and using many fewer vehicles than we did, were less troubled by these factors. Often, in finding an area that was accessible and level enough for a regiment to spend a night in, we would discover that the Germans had occupied that very site and deliberately – with "malice aforethought," as lawyers would say – before withdrawing had defecated at random. Little heaps of human ordure were fresh and ready to be stepped into or slept upon if one was not careful.

The stench of this disgusting refuse, combined with that of rotting human and animal flesh from the shallow burial of their slain comrades or mules killed by our fire, was sickening. Late one day, as I drove off a road to spend the night with my advanced HQ, we were welcomed by an unburied arm sticking up from the ground, still attached to the German cadaver, scarcely covered by a layer of dusty earth. In the burning Sicilian sun, human flesh became scorched and foul within twenty-four hours. I am thankful to say that we, with our prisoners and burial units, took care of our own dead, and the enemy's also, with no delay, and with the proper humane and decent methods and reverence to which we were accustomed. It was a relief to rest in the orange groves near Mount Etna and to move a few days later to the Militello-Scordia area, where there were no orange groves but plenty of olive and almond trees.

Our own sanitary and feeding and other administrative and supporting arrangements deserve a few words. Each unit had a pioneer and sanitary section. The pioneers did small repairs and always had on hand a few wooden crosses. When a burial was necessary, particulars of the victim, whether friend or foe, were taken from his identification discs (one detached for records and one left on the body for future identification) on wrist or neck and painted on the cross. A burial party with the appropriate padre (Protestant, Roman Catholic, or Jewish) dug a proper grave, wrapped the deceased in a blanket, took off his boots (because the thrifty Sicilians were short of leather and would be tempted to partly disinter the body and remove the soldier's boots), performed the usual service, covered the grave, and fixed the cross firmly at the head. When the campaign was over, all these graves would be opened and the bodies taken to a central cemetery where proper identifying headstones would be erected.

So they died. How did we live? Food and water were essentials. When we landed we carried in our kit emergency hard rations of tinned beef (bully), biscuits, tea, and water. A day or two later the "campo packs" arrived. These were wooden boxes containing composite rations for fourteen men for twenty-four hours and included cigarettes, toilet paper, pepper, salt, usually a condiment, tinned meat and vegetables that could be eaten as they came or warmed up over a sterno stove or a makeshift stove of a tin of sand with a cup or two of petrol thrown in. There were a variety of campo packs: "A" would have tinned stew, "B" tinned bacon, "C" bully beef, "D" mutton stew, and so on, and there were a variety of tinned fruits. Obviously some types of campo packs were more in demand than others.

My good man Baker worked out a good scheme. He took an empty metal mortar ammunition box and fastened it in front of us on the hood of the jeep. The windshield was always kept lowered and covered with canvas so that it would not reflect the sun's rays. In this ammo box, he always kept two or three days' supply of food and managed usually to scrounge the more tasty varieties, such as tinned peaches, figs, and Fray Bentos Beef (the best). A jerry can of water from a regimental water truck was always at hand in the jeep, so that he could produce a mug of tea or a reasonable meal in short order. This was the way I lived when I was often away from my main Brigade HQ for twenty-four hours or more. But I was always in touch by wireless.

We were particular about sanitation, and within minutes of stopping for a few hours' rest, every unit had one or more open latrines dug with a rail or bar across the top (for comfort and safety!) and a urinal dug nearby. We were particular, too, about the use of the mosquito nets and about taking the little yellow mepacrine pills to guard against malaria. The drill was to dig with our small entrenching tools shallow slit trenches two feet deep in which to lay our weary bones. Many a man owes his life to this simple expedient defence against enemy air or artillery shells.

Water was scarce and carefully husbanded. It came to us in water trucks and was doled out in petrol cans and thence to water bottles. I recall getting a little lecture from Monty on the subject of water one day. He was sitting with Simonds and Vokes and me under an olive tree waiting for Penhale to arrive for a "briefing on the situation" as it then was. The Americans were on our left and the British 78th Division on our right. To Vokes and me he said, "How are your men getting along for water? Any shortage?" We both said they were making do but had to be very careful. "Good, good," Monty said. (He had a habit of repeating words.) "You know, Englishmen bathe too much, too much. They think they need a bath every day, just nonsense, nonsense. This dust isn't dirty." He brushed his trouser leg. "It looks dirty but it just blows off. All I need is a mug of water in the morning, just a mug, first clean my teeth. Doesn't take much, then shave. Doesn't use much. Then wet a sponge and wipe my face, and the mug has enough left for a sponge bath. Plenty. Englishmen waste water, waste it."

Monty asked another question about the quality and supply of food, which we assured him was first-class. It gave rise to an interesting anecdote about his experience with King George VI when he came out to North Africa shortly after the campaign there was over. Monty's story went something like this.

"When the King came out to see us after we finished off the Germans, I put him up in my visitors' caravan, very comfortable it was. I knew he had trouble with his stomach, you know, so I saw to it that he got the right kind of food and all was well. But then after two or three days, he decided he would like to visit Malta, so we flew him over there. They didn't know about his stomach trouble and fed him the wrong kind of grub, so when he came back to me he had a bad stomach ache, and was very crotchety. Too bad. Nothing seemed to please the old boy. I was going to take him up to Tunis to see some of the units and intended to use my open car, but he disagreed, said he wanted a closed car. So off we went in a closed car and soon had problems. You know, he has a speech impediment and had to sort of get all charged up, and then let go with a sentence. Well, the first regiment we stopped at, the CO came to the car window and saluted smartly. I rolled down the window and told the King what unit it was. He took a moment to get all charged up, and then let go with a stock question. 'How long have you been out here?' The CO couldn't hear him, as I knew would be the case, so he said, 'I beg your pardon, sir?' Well, that meant an embarrassing long pause so the Old Man" – a term of respect in the Army – "could get charged up again, so I just said loud and clear, 'He's been out more than two years, sir.' I didn't know how long he had been out, but it didn't matter and it saved an awkward pause. I knew this sort of thing would happen when he insisted on using the closed car and talking through a half-open window."

On my brigade staff I had a Roman Catholic padre, Captain Reverend A.J. Barker, and he served the Catholics in all three regiments and attached troops. "A.J.," as we called him, being able to converse in Latin, was able to pick up the Italian language quickly, and though I had an English officer fluent in Italian who questioned prisoners and was my official interpreter in dealing with civilians, A.J. was most helpful in scrounging little extra treats, such as fresh fruit and some vegetables from the natives. While we were at rest near Militello awaiting the move to Italy, he became friendly with the sisters in a convent there. They, having seen the little triangular pennant which I flew on my jeep to indicate that I was the commander of the 1st Canadian Infantry Brigade, made a large duplicate in the same colour and design and sent it to me by A.J. Thereafter this pennant flew at my Brigade HQ through the Italian campaign. I still have it, though now much faded and somewhat mutilated by sun, wind, and occasional bits of hot shrapnel! It is a pleasant reminder of not only the kindness and gratitude of the reverend sisters in Militello, but also of the friendly reception we got from the Sicilian and Italian people as we drove their hated allies from their domain.

During our few weeks in the Militello area we were not so idle but kept at a program of training, physical exercises, and making, mending, and refitting. To make it a bit more bearable because of the intense heat, our hours of work were from 4:00 to 8:00 a.m. and 5:00 to 7:00 p.m., and even then, I scarcely need to say, the work was not strenuous. The early morning hours were quite pleasant, and it was a pity to spend them in sleep. One Sunday, during our rest period, I had the pleasure of receiving General McNaughton. I had heard rumours that during the campaign in Sicily he had come out to Malta intending to come across and see his Canadians in active operations, but Monty had forbidden this, and he had to return to Britain. Now that the fighting was over, he was permitted by the Army Commander to make his visit to see us. As I recall it, he was not accompanied by Simonds. It was a Sunday, and he had lunch with my Brigade HQ officers and me and then visited the various units informally.

Andy McNaughton was, I think, a scientist more than a soldier, and his questions to me were more with reference to how our equipment performed than how the men survived the heat, lack of water, rough terrain, and tenacious enemy. This is not to mean that he did not appreciate the skill and valour of his Canadian troops. A few years later in London, when I went to pay a courtesy call upon Monty when he was CIGS, he asked me, "How is Andy McNaughton these days?" I said he was very well and keeping busy, whereupon Monty said, "You know, Andy doesn't like me because I wouldn't let him go to Sicily. It was Guy Simonds who didn't want him. Guy said, 'For God's sake, keep McNaughton away, I don't want him looking over my shoulder,' so I had to tell Andy he would have to wait until the fighting was over." If McNaughton seemed extremely cool to Guy Simonds, no doubt the above episode was one of the reasons.

The next two weeks passed pleasantly and all too quickly. Regimental and brigade sports meets were held. The RCR won the brigade champion-ship and received a cup contrived by the armourer by using a brass shell case and fixing it with rather attractive handles and a wooden base. Swimming trips were arranged, and trucks took the boys to the excellent beaches not far distant. My Brigade HQ was sited at the edge of a ravine, and early one morning a sentry of my defence platoon spied an elderly native leading a mule on which was a mattress and astride this a young and (according to the sentry) comely wench, apparently out to make a few honest lira by practising the oldest of professions. The sentry reported that he sent the old retainer and his handsome baggage back the way they had come. There was only one witness to this episode, but the sentry seemed honest and straightforward in his telling thereof, and I had no reason to doubt it!

10

ASSAULT ON ITALY

Late in August we were called to Division HQ near Scordia, not far from our site, and were briefed on the plan for the impending invasion of Italy. The date was to be the night of September 2, 1943. The 3rd Brigade, composed of the Royal 22nd Regiment, the Carlton and York of New Brunswick, and the West Nova Scotia Regiment, was to make the initial assault on and clear the city of Reggio di Calabria at the very tip of Italy. My brigade was then to pass through Reggio and clear the forts and high ground above the city. Further exploitation inland was to be the task of the 2nd Brigade.

The Calabrian peninsula, which forms the toe and instep of the Italian boot, has been described as a citadel of granite, and so it proved to be. The Apennines, extending from the Alps to the Strait of Messina, dominate the long and narrow land mass that constitutes the boot of Italy. Where the Apennine Mountains do not fall steeply and abruptly to the sea, there are small areas or valleys of excellent fertile land, with long sandy beaches in many places. It is in these locations that the coastal cities and towns are located, and behind them the mountains rise in a succession of plateaus and terraces to their rugged peaks some five or six thousand feet above the sea. On the lower slopes, however, can be found groves of orange, olive, and lemon trees and prolific vineyards. In the winter and spring seasons, abundant snow can be encountered at the higher levels, and when this is melting and heavy rains prevail on the lower slopes, the water rushes down in very broad but shallow beds in terrifying torrents. This was the type of country we were facing in the sunny, warm days of early September, 1943.

In the Foreword to his memoirs, Field Marshal Alexander wrote: "I was concerned that the allied campaign in Italy should receive the recognition in history for its contribution to the general victory in the west. Strategic considerations apart, the seemingly unending succession of mountain ranges, ravines and rivers of the Italian terrain demanded the soldierly qualities of fighting, valour, and endurance in a measure unsurpassed in any other theatre of war. The reader will learn what reflections still vividly remain on the horizon of memory, for the officers and men who fought over them."

Late in August, in Sicily, my orders were to concentrate the brigade in an area about twenty miles south of Messina near the town of Santa Teresa, where we would be readily available for embarkation in LCAS, LCTS, and LCMS, which were scheduled to appear on the beaches in that area the evening of September 2. Therefore, early one morning, with Baker in my jeep, we drove up the coast road through Catania to do a reconnaissance of the 1st Brigade's designated area. Our road ran close to the Adriatic Sea for most of the way and it was a pleasant drive. Just south of Santa Teresa, a bridge, which had crossed a very broad *fiumiari* or shallow riverbed, had been blown. Engineers were sweeping this dry riverbed with mine detectors and taping with white ribbon a safe, broad passageway for vehicles so there would be no serious delay in moving large convoys of vehicles straight on to Messina. As we drove slowly between the swept and taped roadway in the riverbed, I saw that it broadened out upstream, with a number of dried streambeds or tributaries leading into the hills. It looked to me like an ideal place to hide my entire brigade preparatory to crossing into Italy.

I told Baker to turn left and we would go and have a look. Stupid ass! As soon as we turned off the taped area, there was a roar from the engineers working nearby, and one of them came running up to us when we had stopped. He shouted, "Where the hell are you going? This place hasn't been swept. We know there are mines in it because a horse was blown up this morning. Look here!" He stooped over and scratched loose gravel a foot from the front right wheel of the jeep, and sure enough there was the fuse of a mine. I was chagrined and explained that I was looking for a place to put a brigade for a day or two and thought that the tributary streambeds looked suitable. I asked if he and his men could go ahead of me and see if there was a clear way. He was kind enough to agree, and though we saw the dead horse and damaged cart we found no more mines. The brigade in due course was able to tuck itself away in the lemon groves that grew along the banks of the main river and its tributaries. Being early September, the lemons were fully ripe and could be eaten almost like oranges. By evening, Baker and I were back at Brigade HQ, and by August 31 the whole brigade had moved to the new area preparatory for embarking in the landing craft that would take us across the Messina Strait.

Messages were received from General Montgomery and General Simonds and were read to the troops. Montgomery wrote: "I want to tell all of you, soldiers of the Eight Army, that I have complete confidence in the successful outcome of the operations we are now going to carry out. We have a good plan and air support on a greater scale than we have ever had before. There can only be one end to this next battle and that is ANOTHER SUCCESS." Simonds wrote: "1st Canadian Division is in the van of this invasion of Continental Europe. It is a grand honour for this Division, earned by virtue of its achievements in the conquest of Sicily and it is an honour to our country that Canadian troops share in this decisive act."

During the night of September 2, I recall sitting on a stone parapet, along the road just south of Messina, with Vokes and some of our brigade staff and hearing and watching the shattering blast of high explosive shells fired from ships of the Royal Navy as they fell on the city of Reggio and the hills above. This was the softening-up program preceding the landing of the 3rd Brigade on the beaches of Italy in front of and south of Reggio. The division plan called for the 3rd Brigade to lead the assault and clear Reggio, followed by the 1st Brigade, to go through Reggio and into the hills behind to clear anticipated strong enemy positions. When these tasks were accomplished, the 2nd Brigade would pass through and exploit into the high ground of the Aspromonte plateau to the limit of their resources. All went according to plan, even better than anticipated. There was very little enemy resistance, practically no Germans, and perhaps the worst problem was handling the great number of Italians who were delighted with an opportunity to surrender. The only real excitement, and it was short-lived, was just as the landing craft in which I was crossing the Strait touched down on the sandy beach at Reggio, a half dozen enemy light fighter bomber planes flew in at a low altitude and dropped their bombs. They fell into the sides of hills behind the city, where they did no harm.

There is no need to continue to describe in detail our advance up the boot of Italy. An announcement was made on September 8 that Marshal Pietro Badoglio, representing the Italian government, had signed unconditional terms of surrender. Thereafter the Italians surrendered in droves and, with rare exceptions, were anxious to give us all possible assistance to demonstrate their hatred of the Tedesci, as they called the Germans. By September 9, the 1st Brigade had advanced to the city of Catanzaro, the narrowest part of Italy, only twenty miles from the east to the west coast. Here we had a few days rest to "pull up the tail," as Monty called it, because we had outrun our administrative services. However, by September 16, we had moved again with little or no resistance about seventy miles farther north to the town of Castrovillari. This was a week after the assault landings at Salerno just south of Naples by British and American forces and they were having a very difficult time. The 5th British Division had been expected by this date to have advanced far enough north along the west coast of Italy to support this assault on the beaches at Salerno.

However, the 5th Division had been stalled by enemy delaying actions, and the assault forces at Salerno were in need of support or threats upon the enemy from the east, i.e., from the area of Potenza. The 1st Canadian Division was ordered, therefore, to take that city and the utmost speed was essential. By road, Potenza was about 150 miles north of the 1st Division's location. The 3rd Brigade, under Penhale, was given the task, and it is worthy of note that elements of that brigade entered Potenza during the night of September 19-20, having brushed aside pockets of enemy resistance and overcome serious road craters and mines along narrow, difficult mountain roads. The importance of this speed and success of 3rd Brigade's operation was summed up in this extract from a short Canadian government document: "The sudden appearance of the Canadians on their [German] left flank apparently convinced the Germans, still grouped menacingly around the Salerno bridgehead, that the game was up." General Dempsey, the British 13th Corps Commander, wrote to General Simonds: "I hope you realize what a great achievement the capture of Potenza in sixteen days (from day of landing) has been and what a very big effect it has had on the Salerno operation. I offer you and your Division my very sincere congratulations." I think it is indeed a little-known fact that it was the Canadian Division, from its capture of

Potenza and the threat it posed to the German rear, that resulted in final success at Salerno. The long and rapid movement from Reggio to Potenza of 375 miles by road, through natural barriers made more formidable by the enemy's lavish use of high explosives, was a memorable if relatively bloodless operation of war.

A second element in the Potenza plan involved an independent operation by the 1st Canadian Brigade that took it far to the north along the coast of the Gulf of Taranto. It was my task to establish a firm base to protect the right flank of the 3rd Brigade on its advance to Potenza. I was to join hands with the 1st British Airborne Division, which was landing at Taranto, and patrol boldly to the northwest. It was vitally important to conceal from the enemy the fact that only one division was operating on the 150-mile front between Potenza and the Taranto area. Great care had to be taken not to lose any of our men as prisoners. My most active element, in addition to my three battalions, was the 4th Reconnaissance Regiment (The Princess Louise's Dragoon Guards of Ottawa). This unit roamed as much as a hundred miles from its base near my HQ and fought a number of small but fierce patrol actions with elements of the German 1st Parachute Division without serious casualties.

During the three weeks the 1st Brigade was on its own, so to speak, I was out of radio communication with the divisional commander but had an unexpected visit from Lieutenant-General Alfrey, the commander of 5th British Corps in whose area I had been. He flew in by a Cessna recce plane and stayed about an hour. It was the first time I had met him. He seemed quite satisfied with what was being done, and my next contact with him was when we came under his 5th Corps command at the Moro River. In fact, these three weeks were very pleasant. We were near the sea, the weather was fair, and though we did patrol vigorously and far afield, it was the most pleasant period of my entire time in the Mediterranean theatre.

On rejoining the division near Foggia on September 30, I was ordered to move due westward into the mountains, with my final destination the city of Campobasso some eighty miles distant. The next two weeks saw heavy engagements with the enemy, always able and very proficient in selecting defensive positions, as they had done in Sicily. There we suffered from intense heat and dust; now we encountered heavy rains and gooey Italian mud, and streams that were beginning to swell and thus forbid easy fording across the shallow, dried, gravelly beds we had experienced a few weeks previously. I cannot emphasize too strongly the tenacity, fortitude, and valour of all units in their struggle across a distance of nearly one hundred miles through the Apennines. During these two weeks, we were supported to the limit of their endurance by three regiments of tanks, the divisional artillery, engineers, signals, medical services, and other administrative units.

Early on the morning of October 12, I went forward as usual to see how the troops were doing. For a change, it was a lovely autumn day. I met Major Kennedy, then second-in-command of the Hastings, and walked along a rather wide and pleasant valley for a few hundred yards to where his forward company was located. They would shortly be moving on toward Campobasso. We walked back the way we had come. I bade him good hunting, and with Baker, who had been waiting with the jeep, I made a few more calls and was back at my HQ by noon. Shortly after, I had a visit from Lieutenant-General Miles Dempsey, the 13th Corps Commander. We knew each other slightly because for a time in England, as a Brigadier on the General Staff, he had been attached to the 1st Canadian Corps when it was being formed. "Bimbo," as we called him, was a very friendly type, and after a few words he said, "Graham, I am most anxious to get Campobasso by the 15th [October] and the Divisional Commander says your brigade is detailed for the job. Do you think you can make it?"

"I am quite sure we can make it by the 14th," I replied, "unless Jerry makes a heavy counterattack, and I don't think that will happen."

"Good," he said, "if you have cleared the place by the 15th, I'll bring you a bottle of Marsala." On the 14th the RCR went through the city of 15,000 in an advance that their CO, Lieutenant-Colonel D.C. Spry, said was "absolutely bloodless." And Bimbo was as good as his word. A few days later he called at my HQ and left the bottle of sweet Sicilian wine. But sad to say, I did not taste a drop. It was left on a table in my caravan, and when it was moved shortly after, the bottle fell to the floor and smashed.

After we parted on the morning of the 12th, and Major Bert Kennedy was making his way to locate the RCR on his flank, he was ambushed, not far from where I had left him, and was taken prisoner. I received news of this late on the 12th, and knowing Bert as I did (he had been one of my company commanders in the Hastings), I recall saying to my brigade major, "I'd bet even money Bert will be back with us before long." And sure enough, at dusk a day or two after his capture, he was being taken in a truck with other prisoners along a mountain road toward Rome. As they were travelling along the side of a wooded ravine, Bert, having placed himself at the back of the vehicle, tumbled over the tail-board, and in a jiffy was in the ravine and out of sight. The story of Bert's next three weeks deserves a chapter all its own. Befriended by Italian peasants, but knowing naught of the language, dressed in the discarded ragged garments of an old farmer, he wandered, dodged, and tramped the mountains east of Rome until he reached U.S. Army posts about fifty miles west of the Canadian Division front line. The Americans drove him back to his regiment where he turned up in his tattered garb on November 15. A day or two later Tweedsmuir was evacuated with jaundice and Kennedy took command of the regiment once again. Tweedsmuir, after recovering, was posted to the Staff of Corps or Army, and was not again connected with the Plough Jockeys until after the war, when he served a term as the Honorary Colonel.

Once Campobasso was cleared, it was still a vital necessity to drive the enemy some fifteen miles to the north and west so that they could not continue to shell and harass that large and attractive city which General Dempsey intended to use as a leave and recreation centre for troops of 13 Corps. To accomplish this task of advancing further north was easier said than done. The Biforno River, about five miles northwest of the city, was not yet in full spate, but all bridges had been blown and enough rain had fallen to make the crossing by our assault troops a difficult and costly operation in the face of the still stubborn and determined Germans. However, by October 27, the brigade had completed the tasks set for it and the enemy had been driven beyond the artillery range of Campobasso. My Brigade HQ I set up on a reverse slope north of Castropignano, and there it remained, a collection of vehicles and tents, until near the end of November when the division was moved to the Adriatic coast to relieve the 78th British Division. We were to advance northward along the coastal road and cross the Sangro River. A few miles farther we were to cross the Moro River and thence advance through Ortona to the town of Pescara at the eastern end of the main cross-country road from Rome. This last objective, alas, was never to be reached.

During the month of November, all was not always quiet. Enemy guns could still reach some of our forward positions, and movement during daylight was always dangerous in the forward areas. One day my padre, A.J. Barker, received a message from the Hastings that two of his communicants had been killed and he was required at Molise for the committal service. Since he could not go up to the position in daylight, I said Baker and I would drive him up in my jeep that night, which we did. On arrival at the Hastings HQ, we were guided to the house where the CO and his adjutant and other HQ officers were located. They were sitting comfortably before a roaring fire, and while the padre was taken to perform the burial rites, I had a bit of *vino* and half an hour of relaxation. I mention this because two of those present, the CO, Lord Tweedsmuir, and the intelligence officer, Farley Mowat, were, even then, great storytellers and are now internationally known writers.

On November 17, General Vokes, commanding the division, called an orders group at his HQ in Campobasso and we were given detailed orders for the move that would take us to the Adriatic coast and thence advance to the north as the forward division of the 5th Corps. The move began on December 1, and words fail me in trying to describe the congestion of traffic as our division, including troop-carrying trucks, tanks, guns, armoured carriers, and administration vehicles, made their way, or tried to, forward across the bed of the Sangro. Thence on northward we slowly progressed, to take forward positions from the 78th British Division whose transport was moving south on the same roads and at the same time as we were moving north. It was then, and continues to be, a marvel to me that both the 78th British and 1st Canadian Divisions were not practically destroyed as fighting forces, as they could have been by a very few enemy aircraft. We must thank the valiant efforts of the Desert Air Force, which kept pace with the ground forces, often landing along the beaches of the Adriatic. Finally, by December 4 and 5, I was in a position to take over the right flank of the 1st Division area from the Irish Brigade of the 78th Division. I recall clearly looking from the high south bank of the Moro River across the wide streambed to the heights beyond. In the distance were the high cliffs on which stood the town of Ortona.

Major-General Vokes's plan was to make the axis of his main thrust across the Moro River by a good inland road that ran from the village of San Vito on the high ground to the south to the village of San Leonardo on the north bank. In order to distract the enemy's attention from this plan of attack, I was directed to make a feint attack late on December 5 across the broad but not yet fully flooded mouth of the Moro. This task I assigned to my right forward battalion, the Hastings. The feint attack proved successful in that it provoked enemy counterattacks with infantry and armour and no doubt left them with the impression that the main thrust toward Ortona would be by my coast road. Nevertheless, the Hastings attack, though intended as a feint, had revealed the fact that a bridgehead across the Moro at that point was probably feasible and not too costly. With the consent of the division commander, I had the Hastings do just that. By manhandling anti-tank guns across the riverbed, they succeeded in driving off enemy counterattacks supported by tanks.

I went across to see the regiment and at the same time get reliable information by a close view of the ground and by picking the brains of Kennedy, then commanding the Hastings, and his officers. A long time afterward it was rather rewarding to read in a short history of those days these words: "The Hastings were greatly heartened by the presence of their old Commanding Officer, Brigadier Graham, who came forward and stayed with the advanced Platoons until the immediate danger was past."

But the Hastings was not my only unit to bring great credit to themselves and the brigade in this operation. The 48th had forced a crossing on the river left of San Leonardo almost three miles to the west of the Hastings bridgehead. It was impossible for me to get across to see them, as I was trying to plan a move by the RCR whereby they would assemble in the shelter of the Hastings bridgehead and from there move across the enemy front and join hands with either the 48th or units of the 2nd Brigade that might force a crossing at San Leonardo. Such a move across an enemy's front was one of the most hazardous in the realm of tactics. The danger of an enemy attack against the north flank of the RCR was a risk I took. I had hoped that with the Hastings as a base at the coast and the 48th and 2nd Brigade units in or near San Leonardo, the enemy would be loath to commit much strength against the RCR. I was wrong. One company of the RCR did reach the town of San Leonardo, but the enemy counterattack with tanks, mortars, and machine-gun fire was so intense that Lieutenant-Colonel Spry, CO of the RCR, wisely withdrew and reorganized on the south bank of the Moro. I now quote from Nicholson's official history:

As day broke on the 9th, the enemy still occupied San Leonardo and dominated the river below. But now armour could cross the Moro. At 4:30 a.m. Brigadier Graham had signalled General Vokes, "It appears that it is not possible for me to form the bridgehead as ordered by you, but at least the operation enabled the division to be prepared. . . . It would be of great assistance if tanks were pushed over as soon as possible. . . . It has been an exceedingly busy night." On the previous evening, after waiting in vain for a success signal from across the river, sappers of the 3rd Field Company

R.C.E. had begun constructing a diversion around the blown bridge on the main axis, coming under harassing shellfire and sniping in the riverbed. One of the heroes of the night was a bulldozer operator, Sapper M.C. McNaughton, who drove his cumbersome machine across the exposed river flat, making "as much noise as an entire tank brigade" and "under continual machine-gun, mortar and shellfire . . . quickly and skilfully cut down the far bank." At 6:00 a.m. the diversion was ready for use.

Unfortunately the Sappers, who had survived the hours of darkness with very light casualties, were caught in a barrage which met the Canadian armour as it began to move across at seven o'clock. One man was killed and 21 wounded. McNaughton's contribution to the success of the operation was typical of the fine support given by the Engineers throughout the campaign, and won him the Military Medal. . . . When the fighting on the 9th was over the diarist at Cdn. Div. H.Q. could well write, "This day will be remembered by the 1st Canadians for a long, long time. We had our first real battle on a divisional level with the Germans." In the evening General Montgomery signalled, "Hearty congratulations on day's work and on throwing back counter attack." The Hastings were left in their bridgehead near the coast, possession of which they had undeniably established in three days fighting.

A recurring medical complaint now took its toll. In 1939-40, I was diagnosed as suffering from a duodenal ulcer. A few days' rest and a careful diet made me fit again for active duty. It struck again in early 1942, but after a few weeks the little devil healed. It took this opportunity to strike again. I could not eat, I could not sleep, I lost weight, and with the help of my staff captain and Baker, who scrounged for bland food, I did my best to find the relief I had gained in hospital in 1940 and 1942, but without success. At last on December 16, Colonel Boyd, commanding the casualty clearing station for my brigade, said I must go back to No. 1 Canadian General Hospital, then located at Bari on the Adriatic coast, for x-rays and diagnosis. Baker would drive me the 150 miles and stay with me until I was well enough to return to the brigade. I called at Divisional HQ and told my good friend and now GOC, Chris Vokes, of my problem and my hope that, as on the previous occasions, I would recover after a couple of weeks' rest and treatment.

But this was not to be. At Bari, following the usual barium diet and x-rays, the ulcer condition was confirmed and I was put to bed. Christmas Day the nurses came through the ward and sang carols and tried to make us, the patients, as happy as possible. No turkey and plum pudding for me! On New Year's Day 1944, I was transferred by ambulance across the boot of Italy to No. 14 Canadian General Hospital, located in a cavalry barracks at Caserta about eighteen miles north of Naples. Here I shared a room with Lieutenant-Colonel H.M. Hague, who had commanded the 2nd Field Regiment, RCA, which had supported my brigade. We called him "Gen," short for "General," a nickname he had acquired while serving in World War I. Though spelled differently, his name was pronounced the same as Douglas Haig, the British Commander in Chief! While at my headquarters south of the Moro, Gen was with me in a small dugout when we were being shelled late one evening in early December. Suddenly a soldier raised the blanket at the dugout entrance and shouted, "Colonel Hague, one of your men has been hit." Without a thought for his own safety, Gen rushed out to see if he could help his wounded man and was at once hit on his right arm by flying shrapnel. He lost his right hand above the wrist and, like me, his fighting days were over. I followed him to hospital a few days later, and we travelled to Caserta and finally to England together. Gen and his regiment gave us every possible support, and to this day it is a great pleasure to hear periodically from my old friend and comrade, living in Westmount, a borough of Montreal.

I have forgotten how many days it took to steam from Naples to Bristol, but it was quite an experience to be in a ship ablaze with light and with an enormous red cross painted on its sides. As a matter of fact, I had not seen so much artificial light since I had left Halifax more than four years earlier. On arriving at Bristol, hospital trains took us to our various destinations, depending on what treatment we needed. I was taken to No. 10 Canadian General, north of London. Here I remained for some weeks, feeling quite fit but still on an unexciting diet and having x-rays periodically taken of my innards. I was not made for a lazy life so was relieved and glad when word came at last that I was able to travel and, as soon as transportation was available, would be returned to Canada for a staff appointment at Defence Headquarters in Ottawa. I had not long to wait and the transportation was rather unusual. I was given a railway warrant to Kilmarnock, Scotland, and told I would be met there and taken to Prestwick for an Air Force flight to Montreal. All went as arranged, and in due course I was taken to the aircraft for takeoff at approximately 11:00 a.m., having left London the night before. The aircraft was a four-engined Lancaster bomber. There were just four passengers and just four seats, two on each side of the narrow fusilage facing each other. We were given a box lunch (not on my diet) and thermos of coffee. I understood that we would probably land at Gander to refuel, but luckily we had a tail wind most of the way and flew non-stop to Montreal. I was told that this was a record crossing of two minutes under thirteen hours.

So ended the period from our mobilization on September 1, 1939, to March, 1944, of which almost four and a half years had been spent in foreign fields, away from home and family.

11

At Army Headquarters

On landing at Dorval, we four passengers were met by an officer with a car, and we started for the city where rooms had been booked for us that night. It was now dark and a dismally cold, wet, early March evening. It was a treat for all four of us to see again the bright lights, and my thoughts were with my Jean and Peter, whom I would see within a few hours, after an absence of more than four years. Scarcely had we gotten clear of the airport when from behind us came the shrieking of a siren and the flashing of a light. Our driver pulled to the roadside thinking it was an ambulance. A motorcycle officer pulled up beside the driver (we were in a khaki-coloured car as used by the military) and demanded to see his licence. The keeper of the peace said we had been exceeding the speed limit and our poor young driver was given a ticket. I had thought we were travelling at a reasonable speed, other cars had passed us, but there was no use arguing. He permitted us to proceed and I told the driver that if he had any trouble about the summons to let me know, and I might be able to help him in some way. I never did hear from him, so I assume that the matter was settled in some satisfactory way.

The officer who had met us was very helpful. He got me a ticket on the train leaving that night for points west, and by early morning I was back with dear wife and son. I had a week's leave before reporting to Lieutenant-General J.C. Murchie, the Chief of the General Staff at Army

HQ in Ottawa. Only then did I learn that my appointment was to be Deputy Chief of the General Staff in charge of training. I was taking the place of Brigadier Sherwood Lett who, having been wounded while commanding a brigade in the Dieppe raid, had been returned to Canada and appointed DCGS. Now fully recovered, he was returning to command a brigade in the forthcoming invasion of Europe. Sherwood Lett, as I had done, served in World War I, was a lawyer and Militia officer between the wars, and had left his Vancouver law practice to go on active service in World War II. In the operations in Europe in 1944-45, he was again wounded. He returned to his law practice and later became Chief Justice of British Columbia. Had he not died from cancer in that office, he might well have become Governor General of Canada. Few could have filled the office better, and none deserved the distinction more.

So it was in late March, 1944, that I followed this gallant and gifted officer as Deputy Chief in charge of training. Reporting to me were three colonel directors. One was in charge of military subjects like the combatant arms and supporting services. Another was in charge of "trades training," which involved artisans like plumbers, bricklayers, mechanics, and carpenters. The third was responsible for academic instructions. Each of these directors, in turn, had a staff of experienced officers, usually lieutenant-colonels, in the various combatant arms and supporting services that together form an army for which they were responsible. My department, all told, numbered almost three hundred all ranks. Across Canada, from Victoria to Halifax (Newfoundland was not yet a province), were a great number of training centres and military schools. At Army HQ, training policies and programs were developed and updated, and periodic visits were made to the training centres to check on the standards of instruction and the efficiency of instructors. This entailed a good bit of travelling by myself and senior members of my staff. The decision as to whether a man was sufficiently trained and qualified for a particular task, and was ready for dispatch overseas to face the enemy in the field of battle, had to be made at the training centre or school by the commandant and his staff at that place. We, at headquarters, set the standards, but they had to enforce them.

As the months went by, voluntary enlistments decreased, and we were finding it increasingly difficult to meet the demands from overseas for reinforcements required for infantry units in the field. There were many reasons for this situation. After the fall of France in 1940, the Canadian government, realizing that an all-out effort must be made if the Germans were to be defeated, passed the National Resources Mobilization Act (NRMA). This gave the government vast powers over persons and property, including compulsory service in the armed forces. But in this provision of the NRMA there was a very important restriction. The compulsory military service was to be confined to Canada. Men from age twenty-one to age forty-five could be called up for an indefinite period, but only for home defence. However, in 1942 it became evident to the government that if the overseas army was to be increased to five divisions, with two Corps Headquarters and an Army HQ, compulsory service abroad might become necessary, and hence an amendment to the NRMA should be made. With this probability in mind, and to satisfy himself that a great preponderance of the people in Canada would support such action, the Prime Minister with the support of his cabinet in April, 1942, directed that a plebiscite be put to the people. The question was in this form: "Are you in favour of releasing the Government from any obligation arising out of any past commitments restricting the methods of raising men for military service?" The question was answered in the affirmative by an overwhelming majority, Quebec being the only province to register a negative vote.

This plebiscite was held in April, 1942, and the NRMA was amended shortly after that date, giving the government the power to send NRMA men overseas when required. As early as August, 1944, only two months after the Normandy invasion, the Chief-of-Staff at Military HQ in London notified Defence HQ in Ottawa that there was a shortage of 3,000 infantry in the regiments in northwest Europe. Men who had been trained and qualified in Canada to serve in the Artillery, in the Army Service Corps, and in other branches of the Army, were being retrained in England as infantry men to fill existing shortages in that corps. It was forecast that some 25,000 infantry casualties would occur by the end of 1944, with only about 18,000 infantry reinforcements available from all sources, including those retrained from, as mentioned above, other corps. It seemed evident, therefore, in November, 1944, that it was essential to act on the powers vested in the government since the plebiscite in 1942. Trained NRMA soldiers who had remained in Canada for two or more years, serving principally with units in western Canada, should now be sent overseas. This was the opinion of Colonel Ralston, the Defence Minister, who had been in Britain and returned to Canada on October 18 and put his view to the Prime Minister and cabinet the next day. However, Mr. King thought then, as he and many members of his cabinet had thought from the very outset of the war, that "resort to conscription for overseas service would occasion the most serious controversy that could arise in Canada," and indeed they were right! The result was that they did not accept Colonel Ralston's recommendation and decided on another course of action.

Mr. King conceived the idea that General McNaughton might succeed where Ralston had failed in finding the much-needed infantry reinforcements. McNaughton had recently been removed from the Overseas Army Command for reasons of health, or so it was said. Perhaps the General's health had temporarily deteriorated. In fact, he had had violent disagreements with Ralston, his Defence Minister, and with the higher command of the British forces in Europe, particularly Alan Brooke, the Chief of the Imperial General Staff, and Monty. (It was suggested, but less openly, that he had disagreed with Alexander and Churchill himself.) They all seemed to be of the opinion that McNaughton was meant to be a scientist and not a commander of an army in the field, and they agreed with Ralston (and the Canadian government, apparently) that too many Canadian troops had been too long standing idle in Britain while the British and Americans were fighting in the deserts of North Africa and through the hills and mountains and dust and mud of Sicily and Italy. The 1st Canadian Division had arrived in Britain by January, 1940. This was followed by the 2nd, 3rd, 4th, and 5th Armoured Divisions. By August, 1943, only the 1st Division in Sicily had seen active service, except for a part of the 2nd Division in the ill-fated Dieppe operation. All these formations had composed the 1st Canadian Army with McNaughton in command. Naturally he was proud of his troops and referred to them as "a dagger aimed at the heart of Berlin." One can appreciate his opposition to the Canadian government's decision to send the 1st Division to the Mediterranean theatre in July, 1943, and there become part of Montgomery's 8th Army. His opposition was even more violent when he was overruled by Ralston and his cabinet colleagues and the 5th Armoured Division was also sent to the Mediterranean theatre to join the 1st Division. These two divisions formed a corps with General Crerar in command. So it was that McNaughton's Canadian Army was badly split, with two divisions and a Corps HQ in Italy under Alexander's command (Monty had gone to Britain to prepare for the 1945 invasion of Normandy), and three divisions and a Corps HQ Army HQ remaining in England under McNaughton.

The conscription crisis and cabinet's rejection of Ralston's recommendation occurred on October 19, 1944. That night I was sleeping soundly in my bed when the phone rang at my bedside shortly after midnight. "Hello," I mumbled.

"Howard?" I recognized General Murchie's voice.

"Yes, sir," I replied, then wide awake.

"Can you come down right away? We are having a meeting in my office with the Minister and you may be needed."

"I'll be down in fifteen minutes," I said, and hung up. I told my wife I should not be more than an hour, donned a civvy suit, backed the car out, and was in Murchie's office in short order.

There I found Ralston, Major-General Gibson (the Vice Chief of General Staff), the Adjutant-General (Major-General A.E. Walford), and Murchie. Almost at once, behind me the office door burst open and in came Angus Macdonald, MP from Nova Scotia and Minister for the Navy, whom I had never met. He pulled off his rather battered felt hat and flung it on Murchie's desk and in an angry voice said, "Well, what the hell do we do now?" No one had a good answer. Macdonald, I know, had supported Ralston. Both came from Halifax and were lifelong friends. Apparently they had just come from the cabinet meeting where Ralston's proposal to send NRMA men overseas had been rejected, and Colonel Ralston had called this meeting in Murchie's office to tell us of the situation and, perhaps, to confirm that, indeed, there were thousands of fully trained infantry soldiers in Canada, NRMA men whom the government had the power of dispatching overseas. In the general discussion, there was no mention of General McNaughton, so I assumed that Ralston and Macdonald did not know of the Prime Minister's intention to consult with the recently retired General concerning the reinforcement problem. C.G. "Chubby" Power, the Minister for the Air Force, was not present at our meeting for a number of reasons. The Air Force was not short of recruits for overseas service. He was a member of Parliament for Quebec City, and his constituents had opposed the 1942 plebiscite, and he did not agree with Ralston's proposals. The meeting lasted less than an hour, and I think we all went home to a short and fitful sleep, wondering what the solution would be. My own opinion was that the Army in the field must be reduced by disbanding one infantry division or by reducing battalion establishments to three companies instead of four, or that Mr. King and his cabinet must accept the policy which they abhorred and authorize the dispatch of NRMA men overseas.

This was the situation on November 1, when it was announced that

Colonel Ralston had resigned. General McNaughton was appointed to replace him as Minister of Defence the next day. The General had been on leave from the Army but had retired on September 3. Colonel Ralston called a meeting of all senior officers at Army HQ, which I attended, and thanked us for our loyal support. He went on to say that it was a bitter disappointment for him to leave the Ministry of Defence, as he had hoped to see the war through to a successful conclusion. As he concluded his few words of farewell, tears ran down his cheeks. It was a sad day for him and for all of us who, without exception, held him in the highest esteem for his ability, loyalty, and outstanding service to Canada, as both a combatant soldier and a devoted statesman.

Within a week of his appointment as Minister, General McNaughton called a meeting at Defence HQ of all senior officers from across Canada and of his senior staff members. Some of those attending included Major-General Pearkes and Major-General H.N. "Hardy" Ganong (of chocolate fame), both commanding NRMA divisions in western Canada. These gentlemen sat around a long table and at the head sat the new Minister, who was well known and, I think, highly respected. With him were Lieutenant-General Murchie, Major-General Ralph Gibson, the Vice CGS, Major-General Walford, Major-General Hugh Young, the QMG, and the deputy defence minister. Around the wall sat myself and other deputy staff officers. I recall sitting across from Hardy Ganong. The proceedings were not opened with prayer. Perhaps they should have been, for Andy surely needed the help of the Almighty if he was to accomplish what he had apparently told the Prime Minister he thought he could accomplish, i.e., get the needed reinforcements for overseas without invoking the compulsory provisions of the NRMA. After his introductory remarks, in which he said he realized he had taken on a difficult task and hoped he would have the wholehearted support of all commanders, the new Minister then got down to the problem at hand. This was how to get the NRMA men to volunteer for overseas service and thus avoid the type of crisis that, he reminded us, had occurred in 1918 when conscription for overseas service was imposed by a Union government headed by Robert L. Borden. Most of us in the room that morning remembered the riots and open rebellion that had broken out in Quebec in 1917-18 and agreed that a repetition of those events must be avoided if at all possible.

McNaughton expressed the view that surely the required number of men would volunteer if the necessity was made clear to them. An appeal for their support of their fellow Canadians now fighting in Europe would surely move them, or shame them into offering themselves in sufficient numbers to make unnecessary the use of compulsion. As I listened to his appeal and looked at the faces of the men around the table and behind it, I could only describe their reaction as sullen and unimpressed. They all knew and, I think, respected Andy McNaughton. Most of them had served with him in times of peace and war. But I felt certain that they resented his having been to some extent responsible for the dismissal of Colonel Ralston, another man whom they had held in high regard. And, furthermore, he was telling them nothing new. They had already exerted every effort to induce the NRMA men to volunteer. They had exhausted their arguments. And so when he finished his exhortation, there was silence for a minute or two, though it seemed much longer before someone spoke.

Hardy Ganong looked up and down the room, waiting, I suppose, for some more senior and permanent force officer to comment. Hardy almost always had a pipe in his mouth, and the bowl was always tilted to the left at about a forty-five-degree angle, an unusual way to smoke a pipe. Now he took it from his mouth and, no one else saying a word, he said he had a few comments to make. They went something like this. "Sir," he said, addressing the minister.

"Yes, Hardy, what is it?" responded the Minister.

"I am sure you want us to be completely open and frank about this problem, sir," queried Hardy.

"Yes, indeed, I do. What have you to say?" asked the minister.

"Well," replied Hardy, "I feel sure I voice the thoughts and opinions of every one of us around this table when I say that we want to give you every support we can. We have appreciated the seriousness of this problem for several months, and we have done every damn thing it is possible to do to get these men to volunteer. It would be useless for us to go back and try again. It would be a waste of time and we would be leading you astray if we indicated that there was any chance of success. I'm sorry to have to put it so strongly, but I feel I must give you the true picture."

There was no hand-clapping, but there were significant murmurs of approval and sounds of "Hear, hear." Major-General Pearkes sat next to Ganong and as far as I can recall said not a word. Poor Andy! It must have been a shock to have such a dish of ice water cast upon his hopes. A short discussion followed, nothing new evolved, the meeting ended with handshaking all round, and the visiting officers departed for their home stations.

But there was a sequel. The next morning in the newspaper I read that, in an interview with reporters, Andy said that he had a good meeting with the senior officers and that they encouraged him to believe that they would be able to raise the required number of reinforcements voluntarily. I was astonished because, in fact, they had said quite the opposite. Either Andy was deliberately giving the wrong impression or he had been deaf to Hardy Ganong's statement. Before the day was out, signal messages came to Murchie from many of the officers across Canada who had attended the meeting and who had read their morning newspapers. They all expressed surprise at the published reports and asked for a correction by the Minister. One irate telegram from Brigadier MacFarlane, area commander in Winnipeg, tendered his resignation as a protest against what he considered to be the Minister's misleading statement.

These are the events that led to the much-publicized remark of the Prime Minister that there had been a "revolt of the Generals." McNaughton's efforts had produced no more and perhaps even less than Colonel Ralston could have accomplished, but McNaughton's efforts did at least convince the majority of Canadians that every possible avenue had been explored and now conscription for overseas service must be imposed. The matter finally came to a head on the morning of November 22. About 10:00 a.m., Murchie called McNaughton on the phone and said that he and the Adjutant General (Walford) were now convinced that voluntary methods of recruiting could no longer meet the Army requirements. The Minister asked Murchie to put his message in writing and send it over to him. In this memo the CGS said, *inter alia*: "I must now advise you that in my considered opinion the Voluntary system of recruiting through Army channels cannot meet the present problem. The Military members [i.e., the VCGS and AG and QMG] concur in this advice."

The next day the government decided that conscription for overseas service could no longer be deferred. An order-in-council was passed authorizing the dispatch of 16,000 men, and the first group sailed on January 3, 1945. In all, only 12,000 were eventually sent. The records indicate that, of those conscripted troops, there were only 315 casualties, of which 69 were fatal.

A week or two after I arrived in Ottawa and took up my new appointment, I had a phone call from Colonel Willis O'Connor at Rideau Hall.

He was Canadian ADC to the Governor General, The Earl of Athlone. "Howard," he said, "the Governor General would like you to come to lunch tomorrow at twelve noon. It will be stag, just the members of his staff." I had never been to Government House, nor had I met His Excellency, who was a brother of Queen Mary. I gladly accepted and assumed that His Excellency, who was a veteran of the South African War, wanted to hear something about the Sicily-Italy campaign.

The next day, shoes shining, buttons glistening, I arrived at the stroke of twelve, met His Excellency and members of his staff, had a sherry in his study, and then went in to lunch. I have forgotten the menu but I remember it was served on silver plates. I sat next to him and it did not take many minutes to learn the reason for my invitation. Hardly were we seated when my host said, "I say, Graham, what happened between McNaughton and Brooke? They must have had an awful row to make McNaughton leave his Army."

Too bad! I couldn't give the good Earl a clue. I had to say, "I'm sorry, Your Excellency, I have no idea what made the General give up his command, other than poor health. I was in Italy when the General retired." The subject was not raised again, and we continued our luncheon talking about His Excellency's exploits in Africa when he was a young man.

The efficiency of the training program made the headlines in the early spring of 1945. It came about this way. Major Conn Smythe, a veteran of World War I and a widely known sportsman, especially in hockey and later in horseracing, had volunteered for overseas service and was given command of an anti-aircraft battery in France. He was wounded and invalided back to Canada and was a patient in the Chorley Park Military Hospital in Toronto. Major Smythe always did like to make the headlines, and he really did a great job from his hospital bed! He criticized in the strongest terms the standard of training of reinforcements being sent into the front lines of France. It appeared that he was accusing the King government of murder, as it was responsible for sending such ill-prepared Canadians against the enemy.

The reinforcements going to units in the field in 1944 and 1945 lacked the years of training and preparation that had been given the men during those two or three years in England while they were awaiting action in Sicily and Italy in 1943 or in Normandy in 1944. But I was not prepared to accept Major Smythe's wholesale and vituperative condemnation of the training program, expressed in generalities, as is so often the case with criticism. So I suggested to General Murchie that I interview the Major in his hospital bed and try to get clear and definite instances of the cases he was referring to in such a general way. Wisely I think, Murchie said no to my proposal because the Major and the press would be sure to make further capital from my visit and simply exacerbate the issue. The opposition in the Commons made much of Smythe's accusations for a short time, and then the matter died. I did not ignore the criticism, but was not unduly disturbed because I felt that, after all, I had a few years of experience in training and working with troops in all arms of the service, and knew only too well that some of the boys, being poorly motivated, never were and never would be above reproach!

The few times I met Conn Smythe, I found him to be a whining, peevish person, rude and churlish in his manner. Perhaps that was because I was introduced to him as "General Graham" and, unquestionably, Conn did not like generals. But give the devil his due. Conn was a super-patriotic Canadian and supporter of the monarchy. Furthermore, as is so often the case, behind the rude and snarly demeanour, Major Smythe had a kind and generous heart. In particular, he gave much time and money to help crippled children. Some years later he was appointed Honorary Colonel of the Lorne Scots Militia Regiment. I was a guest at one of the mess dinners, and also present was an English Army brigadier who was the Honorary Colonel of the British regiment that was allied to the Lorne Scots.

It was a very proper and formal affair until Colonel Smythe was called on to say a few words. It was just after the new national flag had been approved by Parliament, replacing the red ensign that bore the Union Jack in one quarter and the coat of arms in another. In his not-too-few remarks, Colonel Smythe waxed quite vehement in his castigation of the French Canadians who, he claimed, were responsible for the elimination of the Union Jack from the new flag. What we needed in this country, he expounded, were more people like King Clancy, one of Smythe's favourite hockey players. In the good old days, when playing against the Montreal hockey teams, Clancy would yell, "Come on, you damn peasoupers, I'll show you how we can handle you guys!" This sort of speech was somewhat off-colour for an honorary colonel to make, and I wondered what the impression of the English brigadier was, not that Conn Smythe would give a damn.

After the enemy's surrender in Europe in May, 1945, we began immediately to run-down and disband most of the training centres. In

July, Lieutenant-General Charles Foulkes returned from commanding the 1st Canadian Corps overseas and replaced Lieutenant-General Murchie as Chief of the General Staff in Ottawa. Murchie went to England to take charge of the run-down of establishments there. By May I had served for almost six years and now planned to return to my law practice, which was still being carried on by Alan MacNab and Isabel Hines. However, both the CGS and the Adjutant General said they had appointments in mind for me that they felt I would like. No promises were made but they said that if I decided to remain in the Army, my future should be successful because of my wide range of experience in the Militia, in command and staff appointments during the war, and, of some importance, my civilian background. I was then nearly forty-seven years of age, and because retirement age for senior officers was fifty-five, I would have eight years to add to my pension entitlement, which, when added to my World War I service plus my Militia service (counted at one-quarter of actual time spent in Militia) and my World War II service, would yield a fair amount. Furthermore, I would still have (I hoped) several years of law-related activity after mandatory retirement. So after much thought and discussion with my wife, we decided that I would remain a soldier. I have never regretted that decision.

As work in the General Staff training branch diminished, work in the Adjutant General's branch increased. The heavy task of repatriation and discharge of the tens of thousand of personnel who were overseas, and the discharge of the many thousands who had been employed in the training establishments in Canada and in the two divisions in western Canada, required additional staff. Because of this shift in workload, I relinquished my appointment as Deputy Chief of General Staff and moved over to the "A" Branch and was appointed Deputy Adjutant General. Almost at once, I was appointed chairman of a committee to review applications from officers who desired to remain in the post-war Army. This committee reviewed the applicant's record of service, educational background, age, marriage status, and any other factors that would affect his suitability for continued service in the reorganized peacetime Army. We then made our recommendation to the Army Council, composed of the Chief of the General Staff, the Vice Chief of the GS, the Adjutant General, and the Quartermaster General, for their approval or rejection. Theirs was the final decision.

In addition to reviewing applicants, we also had the task of reviewing applications for service in the Pacific Force against Japan. This was to consist of one infantry division with supporting and administrative troops to operate under United States higher command and to be organized and equipped in accordance with United States tables of organization. Major-General Bert Hoffmeister was appointed to command this Pacific Force. Bert had commanded the Seaforth Highlanders in Italy and later had commanded the 2nd Canadian Infantry Brigade when I had the 1st Brigade. Later he was promoted to major-general and commanded the 5th Canadian Armed Division with great distinction. His choice as commander of our Pacific Force was popular and well merited. However, as events transpired, this Pacific Force was never needed. After the atomic bomb was dropped on Hiroshima on August 6, and another on Nagasaki on August 9, Japan sued for peace on August 10, and on August 14 hostilities ceased. On August 13, our government ordered the Pacific Force to be disbanded. The ensuing months saw the discharge of the personnel who had been accepted for service in the Far East or their transfer to the reorganized and continuing full-time Army.

12

PEACETIME ARMY DUTIES

After the cessation of hostilities in Europe and later in the Far East, my work as chairman of the selection committee took much time in the "A" Branch. I was involved in Ottawa with the preparation of suggested new pay scales for all ranks, the amount of pensions for length of service and war-incurred disabilities, amendments to Rules & Regulations and Orders, and countless other matters relating to personnel. Many of these post-war matters were settled by late summer of 1946 so the work in the "A" Branch was much diminished.

One day the Adjutant General, Major-General E.G. Weeks, called me into his office and said, "Howard, you have been selected to go to London and take over from General Murchie. Jean and your son will go with you, of course, and the posting will likely be for two years. What do you think of it?"

"Well, sir," I replied, "it's a surprise and I'd like to have some idea of what the job will involve."

"I can tell you one thing it involves," went on General Weeks, "and the Minister wants to see you about this. It's getting things cleared up over there. The war has been over for a year and a half and we still have two camps operating, dozens of buildings in and around London that we're paying rent for, and I don't know how many civilian claims unsettled. You'll have lots to do and your legal training should be a help. Can you be ready in a couple of weeks? You'll have to take some household stuff, and perhaps some food because, as you know, things are still pretty scarce over there."

I said I would talk to my wife and let him know the next day how soon we could be ready. That night Jean and I discussed at length this unexpected and major change in our lifestyle and environment, and decided that a minimum of three weeks would be needed to do the many things necessary to prepare for a two-year or more absence from Canada. The next morning General Weeks expressed satisfaction with the three-week period, and arrangements were made for Jean and me and our son Peter, who had to leave Ashbury College in Ottawa in mid-term, to sail from Halifax in early October.

I went to see the Minister, Brooke Claxton, who had succeeded Douglas Abbott as Minister of Defence. Strangely enough, the three of us – Claxton, Abbott, and I – had several things in common. We had all served in the ranks in World War I in France. On discharge, we had all graduated in law. We had all been interested in politics, they being successful Liberals in Montreal and I an unsuccessful Tory in Ontario. Abbott and Claxton were honest, able, dedicated Canadians, but each had a quite different approach to the role of the Minister of Defence.

If Abbott as Defence Minister could not answer a question from the opposition, he would usually say, with a smile, something like, "That's a good question and I am sorry I can't give the Honourable Gentleman an answer at the moment, but I will reply tomorrow." And he always did. Claxton, on the other hand, hated like sin to admit that he did not know the answer to every question put by the opposition. More often than not, he would try to play down the importance of the subject, or take a shot at answering and often missing the mark, causing backfire and perhaps a correction the next day. Maybe this was reflex action from his days as a gunner sergeant major! When I went to see him about going overseas in the fall of 1946, he did not mince words. He thought Murchie was altogether too slow in getting affairs cleared away and perhaps was deliberately stalling in order to prolong his posting to London with the rank of lieutenant-general. On his return to Canada he would retire because of age. During the whole of the war he served at Army HQ in Ottawa, and I found it hard to blame him if he was now making the most of this first and final overseas assignment!

Our preparations for the long move went on apace. I had four large wooden cases made, each to hold about sixteen cubic feet, which we used for bedding, table linen and towels, and spare clothing to last the three of us for the next two years. In addition, I had two large boxes made, each about twelve cubic feet in capacity, which were filled with food. Half of a whole cheese, wrapped in a vinegar-soaked cloth, with instructions to turn it every week, and many varieties of tinned fruit, vegetables, and meat were acquired from and packed by our grocer in Ottawa. Jean did all this buying and packing of clothing and household goods while I continued at the office.

Each morning there came to my desk a file of all messages received during the night. I was particularly interested in one from Murchie, about a week before I was due to leave. It went something like this: "I note Graham coming to London. This may cause problem because of contretemps between him and Simonds." This was the first and only time I ever saw or heard of any reference to this matter until Colonel C.P. Stacey mentioned it many years later. However, neither the AG nor the CGS mentioned the above telegram to me. What reply, if any, was ever sent to Murchie, I know not. My move to London went on as scheduled.

When the war was over and Charles Foulkes had been appointed Chief of the Canadian General Staff, Guy Simonds was sent as a student to the Imperial Defence College in London, and after finishing there he was loaned to the college as an instructor. In this way, both Foulkes and Simonds retained their lieutenant-general's rank. All others of that rank were retired because of age. During the two years I was in London, Jean and I saw Simonds and his charming wife Kay on many occasions and we frequently played bridge together. I can't say that we were close friends, but certainly our association was amiable and pleasant. Murchie's fears proved groundless.

Once in London we began to appreciate the shortages being experienced by the British eighteen months after the cessation of hostilities. I saw General Murchie, whom I held in high regard, and he put me in touch with an officer who was returning to Canada in a day or two and who had a lease on a small two-bedroom furnished apartment in Dolphin Square, a large apartment complex on the banks of the River Thames within sight and sound of Big Ben. We quickly made a deal with the owner, and in a day or two, when our boxes were delivered, we took up residence on the eighth floor of Duncan House. Peter continued his studies and Jean was issued a ration book and registered for grocery shopping at Harrod's and for meat at a nearby butcher's shop. My offices were in the Sun Life Building in Cockspur St. next to Canada House. The whole building was still being occupied as Canadian Military HQ, as it had been for the past six years. Many other buildings in and near London were still occupied by the military and civilian staff, such as the Wives Bureau, the Records Office, and the Stationery Centre with tons of now quite useless forms and related items. There were two camps in Surrey still occupied and administered on a reduced scale, but involving the use of trucks and cars that required garages and mechanics to keep them going.

After a quick inspection of all these holdings, I soon discovered the reason for the long delay in winding up our overseas business. There were hundreds of men who simply did not want to return to Canada and fend for themselves. It would be tiresome to go into a detailed account of how, within six months, all these establishments were closed out and my small staff of about twenty officers and civilian employees – plus a small RCN and RCAF liaison group – were all concentrated in a house at 10 Hill Street in Mayfair, just a few yards from Berkeley Square. The Navy and Air Force, of course, had done their part in reducing overseas staff, as the Army had done. I must admit that a firm hand had to be used, and I am sure I was looked upon with much disfavour by the many officers who sought and obtained interviews with me and put strongly their case for staying longer to wind up this or that unit. But usually they importuned with no success. To empty the Sun Life Building, I gave orders that the top floor must be vacated by a certain date, then the next floor down, and so on, until we were clear of the place and Sun Life was delighted to again be back in their own premises. There were several hundred wives still waiting to join their husbands in Canada. By using pressure in the proper places, we got the Cunard liner *Aquitania* to take the last bevy of beauties from Southampton to Halifax in early spring of 1947. I went down and had lunch on board to bid them bon voyage.

My official title in London was Senior Canadian Army Liaison Officer (SCALO). In addition to winding down the wartime establishments, getting properties vacated, settling claims by civilians, and finding new quarters in Hill Street for both my staff and the Navy and Air Force liaison staff, I attended, as an invited observer, British post-war army exercises on Salisbury Plain and elsewhere and sent reports back to Ottawa. I attended meetings at the War Office with army representatives from countries within the old British Empire. Views were exchanged on such subjects as the organization of post-war forces, weapons being newly developed, training programs, and so on. Again I kept Ottawa informed. Shortly after the war, Belgium, the Netherlands, and Luxembourg formed an economic union, Benelux, and there developed discussions on the possibility of reaching an agreement among the Western European nations for a united defence force and common defence policies. These discussions, which took place in London, were chaired by an Air Vice Marshal of the Royal Air Force. Canada and the United States were invited to have one representative each sit in on the meetings as observers. I was the Canadian representative and kept the High Commissioner, Norman Robertson, informed, and also sent reports to my chief at home, Lieutenant-General Foulkes. I recall Monty, then Chief of the Imperial General Staff, attending one of the meetings and saying with firm conviction that the first essential for the success of such an arrangement was to have one supreme commander for a Western European Defence Force. I am sure he was convinced that he knew the best man for the job! I believe these talks in London on the defence of Western Europe were the precursor to the North Atlantic Treaty Organization (NATO), which was born in New York on April 4, 1949.

Norman Robertson was High Commissioner during my two years in London. Like everyone else who knew him, I had the greatest admiration for his ability, intelligence, and wisdom displayed in his various appointments in the Department of External Affairs. It was a privilege and an education for me to have this association with him in London. Sometimes, when he had other engagements or was not in good health, I would stand in for him at quasi-military functions. One of these was a meeting of the Imperial War Graves Commission in a government building in Whitehall. It was midwinter, and because of a power shortage, electricity was cut off throughout the city from 9:00 to 12:00 a.m. and 2:00 to 4:00 p.m. The meeting of the IWGC was at 2:30 p.m. I arrived at the appointed room and found the long conference table covered with grey army blankets and down the centre about twenty empty beer bottles with flickering candles stuck in their necks. Across one end of the room a very large picture of the projected military cemetery at El Alamein was dimly illuminated by beer bottles and candles in a row along the wainscotting. It was indeed a weird and doleful setting, but perhaps appropriate for a meeting of the IWGC, made even more so by the fact that there was no heat in the room and we sat huddled in our overcoats. Promptly at 2:30 p.m. – punctuality is the courtesy of kings – the chairman arrived – H.R.H. The Duke of Gloucester. There was no kowtowing or waste of time. "Good afternoon, gentlemen," said H.R.H. "The first item on the agenda, as you see, is approval of the minutes of the last meeting. Would someone like to move that they be taken as read – and approved?" Someone did and the chairman proceeded to the next item. I have attended countless meetings, but never one that was more properly and expeditiously conducted than that one. I do not know whether or not the Duke had much knowledge of the business at hand, but he surely knew how to get through an agenda.

If the meeting of the IWGC was a doleful experience, there were other events to make up for it. For a long time I had been a close friend of Milton Gregg, VC. When I was DCGS, he had returned from overseas and was in charge of a training centre at Vernon, B.C. On one of my visits to that centre in late 1944, Colonel Gregg told me in confidence that he had been invited to accept the presidency of the University of New Brunswick, of which Lord Beaverbrook was a great benefactor. Gregg, not being a varsity graduate or scholar, was hesitant about accepting the appointment. I said his great administrative ability was what was required, and he could forget about his lack of an academic background. He accepted the appointment, and when I was in London he came over on university business. This involved seeing Lord Beaverbrook. Jean and I put up Milton and his wife Dorothy in guest rooms at Dolphin Square. One lovely spring afternoon we drove them down to Beaverbrook's country estate at Mickelham, north of Dorking, for tea with His Lordship, whom I had not met before. We had a delightful two- or three-hour visit, sitting on the outdoor terrace that ran along the west side of his large Elizabethan house called Cherkley Court and looked across a valley of the South Downs. His Lordship told us of the day many years before when he and Rudyard Kipling had put their bikes on the train from London to Dorking. On arrival they cycled up the hills to inspect this property, which Beaverbrook had seen advertised and which appealed to him. He painted a word picture of Kipling and himself sitting in the sun on the grassy slope below the terrace admiring the magnificent view across the Downs and of his making the decision to buy the property. He told us, too, of his banana plantation in Jamaica and of his early days in Canada. And then he told us how he had induced his very old friend, Richard Bedford Bennett, former Prime Minister of Canada and now the Rt. Hon. Viscount Bennett of Calgary, to purchase the adjoining estate, called Juniper Hill. A path of a few hundred yards through the nearby woods ran from where we were sitting to Juniper Hill, and so the two old friends saw much of each other.

It was not long after this interesting and enjoyable afternoon that we were shocked by the news that Lord Bennett had died suddenly in his bath from a heart attack. Except for a housekeeper and valet, he was alone. His long-time secretary and supporter since his days as a Calgary lawyer, Miss Alice Miller, who lived in the house, had left only a day or two earlier for a visit to Canada. Norman Robertson was recovering from an operation, so his deputy, Frederick Hudd, and three service liaison officers, one of which was me, were instructed to attend the funeral. We drove down to Juniper Hill from London, and on arrival were led by the butler to the large solarium. There, alone, slowly pacing among the ferns and palms, I met again Lord Beaverbrook. He and the butler were the only two people on the premises. The coffin, with its burden, sat on trestles in a drawing room. No one else was present. Very soon after our arrival the undertaker and six stalwart bearers appeared, hoisted the casket on their shoulders, and, with our little group following, walked down the winding drive to the small, ancient Anglican church, just outside the gates of Juniper Hill. The church was packed to the doors. After the service, which was conducted by the Bishop of Guildford – Bennett had been a great benefactor in the building of the new cathedral on Stag Hill near Guildford – the former Prime Minister of Canada was laid to rest in the churchyard near the gates of his English home. Our little group and a large number of other friends and neighbours returned to the house, where the housekeeper and staff, before going to the service, had made ready a few refreshments. Then we drove back to London in the early evening hours. It seemed a sad and lonely way for a great Canadian to end his days. A few days later a memorial service was held in Westminster Abbey, and again I was detailed to attend. For the first and only time I had a seat in the choir stalls behind the little melodious boys from Westminster School who formed a large part of the male Abbey Choir.

Not long after the memorial service, my wife and I were again in the Abbey. It was a more happy and historic occasion, the wedding of H.R.H. Princess Elizabeth and H.R.H. Prince Philip of Greece. Our good fortune in being invited to this and other such events as the opening of Parliament and Palace garden parties came about by reason of my quasi-diplomatic appointment. Of historical interest and equally impressive was the service in St. Paul's Cathedral to celebrate the King and Queen's twenty-fifth wedding anniversary. The opening of Parliament was another colourful and important event to which we received an invitation.

In the spring of 1947, I received orders from Ottawa to represent the Canadian Army at the unveiling of a memorial cross by Queen Wilhelmina in the Canadian Military Cemetery at Nijmegen in the Netherlands. The Old Queen, as she came to be known, was then under seventy years of age. She had reigned since the age of ten and, in 1948, abdicated in favour of her daughter, Juliana. Pierre Dupuy, our ambassador to Holland at the unveiling ceremony, represented the Canadian government. Princess Juliana, her husband Prince Bernhardt, and other members of the royal family also attended. After the unveiling by Her Majesty, they placed wreaths at the base of the cross. At the foot of each of the hundreds of graves stood a teenage or younger school child with a small bouquet of flowers, and as the Canadian national anthem was played, they placed their bouquets on the graves before them, the resting places of young Canadians who had made the supreme sacrifice in the struggle for the liberation of Holland. The whole ceremony, culminating in this act of reverence and gratitude, was most moving and impressive. I appreciated the honour and privilege of representing the Canadian Army on this occasion, placing the wreath of poppies at the cross.

In the late summer of 1948, I learned from General Foulkes that I was to be promoted to the rank of major-general and posted back to Ottawa as Vice Chief of the General Staff. One final event, apart from farewell parties, that stands clear in my memory was the last of the season's royal garden parties. The High Commissioner saw that we had invitations not only to attend but also to be presented to Their Majesties in the royal enclosure. As it came our turn to step before King George VI, Princess Elizabeth, and Queen Mary, standing on a carpet under a striped canopy, Mr. Robertson made the introduction and said that I was being promoted to major-general and appointed Vice Chief of the Canadian General Staff and would be leaving England in a few weeks. His Majesty shook hands, congratulated me, and wished me well in my new office. Princess Elizabeth, in whose company I would be travelling some twenty years later, and Queen Mary did likewise.

I was fairly familiar with the duties performed by the Vice Chief and found no difficulty in settling into the seat recently vacated by Major-General Churchill Mann, who had retired. I had not hitherto served directly under General Charles Foulkes but we seemed to hit it off quite well together. He did not mind delegating tasks and I never resented accepting them. If I disagreed with some of his ideas in organization or training or officer postings, I did not mind saying so when I had what I thought was a better idea. Sometimes he accepted my suggestions but more often stuck to his own, which was as it should have been. Foulkes, like Simonds, was intensely ambitious. I have always looked upon ambition as a virtue rather than a fault, so long as one does not allow ambition to override the ethics of fair treatment to one's fellows. I certainly cannot point to any case where these two unquestionably able commanders went too far. Foulkes did have one attribute which I think Simonds lacked, an attribute that Field Marshal Wavell once said was essential in a senior commander: he must be able to work amicably with his political masters' wishes. Perhaps he would even curry favour. Foulkes had it, perhaps in excess. Sometimes he allowed his plans to be too much watered down and diluted to retain what was undoubtedly a cordial relationship with the government.

One of my duties as Vice Chief was to be the Canadian Army representative on the Permanent Joint-Board on Defence, which had been created in August, 1940, as agreed by Prime Minister Mackenzie King and U.S. President Franklin Delano Roosevelt at their meeting in Ogdensburg, New York. In 1949-50, the Board consisted of ten members, five from each country. The five members consisted of one representative from each of the Army, the Navy, and the Air Force; one from the Department of External Affairs (or the Secretary of State); and a chairman from each country. For Canada, it was General McNaughton, and for the United States it was General Guy Henry. During the war years, the PJBD was an exceedingly busy and important body. The members discussed and agreed on defence policies and plans, particularly as they applied to North American defences, and made recommendations to their respective governments. After the war, during my term as VCGS, only quarterly meetings were held and these had to do, first, with plans for the run-down and disposal of U.S. bases or camps that had been established in Canada and Newfoundland; second, with the deployment and siting of U.S. or joint U.S. and Canada radar early-warning stations; third, with refuelling arrangements for long-range aircraft, missile-launching sites, types of missiles to be permitted for launch in Canada, and related matters; fourth, with efforts to obtain standardization in weapons, ammunition, and similar products, dispersal of industry, and sharing of defence production. In addition, we visited and witnessed training exercises on both sides of the border, and in general kept each country informed on the organization, strength, and training policies of our respective armed forces.

Our relations then were – and I assume still are – most friendly and co-operative. Canada, in my opinion, benefited much more than did the United States from the work of the PJBD. For instance, on a visit to our Anti-Aircraft Artillery Training School at Picton, Ontario, I mentioned to Major-General Bolte, my American opposite number, that we did not have a certain type of training equipment. "Oh," he said, "we have a number of those that will never be needed. I'll have them sent up to you." Within a few days several cases with the required articles arrived and never a cent was charged. In the United States, we visited such places as West Point and Annapolis, the U.S. Army and Navy training schools.

One memorable visit was to Key West, at the southern tip of Florida. We flew by Royal Canadian Air Force transport to Washington and then by U.S. Air Force craft with our American members to Key West. The next morning we drove to the air strip and boarded a naval blimp. This took us out over the Caribbean Sea many miles until we sighted ships of the U.S. Navy on an exercise about which we had been briefed the night before. A few minutes later we flew over and settled on the flight deck of a U.S. aircraft carrier where many hands gripped the ropes dropped from the blimp and snubbed it to the deck. A collapsible ladder was lowered from the blimp cabin and, in a minute or two, we transferred to the carrier. The blimp was released and disappeared in the bright blue sky. After a tour of the carrier and a short lecture on the operation in which the fleet was engaged, a destroyer was summoned to our ship's side, a rope ladder was lowered from a higher deck, and we transferred to the smaller vessel. No problem, no peril, a calm sea, and ships standing still – so different from my last transfer from a larger to a smaller ship in the Mediterranean only seven years before. In the destroyer we had another tour of the ship, a light lunch, and an update on the exercise. And then one more transfer, this time from the destroyer to a submarine and again by rope ladder, only a few feet this time. Down we went into the bowels of the craft, the hatch was closed, and we sank to periscope depth. The scope was raised, and for some time we moved quietly below the surface, each of us being given an opportunity to use the scope and watch the fleet manoeuvring. Then, in accordance with the exercise plan, we were treated to some very loud bangs on our hull. We were being depth-charged, and in an instant we took a steep dive to what I was told was 450 feet. For a time other charges could be felt, but weakly. We had avoided destruction! Finally we surfaced. We worked our way shoreward after a final half-hour afloat on the narrow deck of the sub. We tied up at Key West Naval Base, where a car awaited to take us back to our quarters. Not many people have ever had such a unique experience of travelling, in the space of ten hours, in a motor car, in the cabin of a blimp, in an aircraft carrier, in a destroyer, and finally in a submarine!

During these trips of the PJBD I saw and talked frequently with General McNaughton. He was always intensely interested in what he saw, particularly in any new equipment and in any move being made toward standardization between Canada and the United States. But above all, he was determined to protect national interests and Canadian sovereignty – "sovrainity," he called it! The General, whether purposely or unwittingly, had his own way of pronouncing some words. When he was the Canadian representative on the United Nations Security Council, the Russian member was Vichinsky, but Andy always referred to him as "Vichitsky." He always called the Canadian Major-General at the head of Canadian Military HQ in London "Montaig." His name was Montague, but apparently Andy considered that if "vague" was pronounced "vaig," then Montague should be Montaig!

McNaughton's concern about protecting Canadian sovereignty reflected the concern of Mackenzie King and, of course, most Canadians. John Swettenham, in his excellent three-volume *McNaughton*, gave the thoughts of both the PM and McNaughton as follows:

The shape that the defence of Canada would assume when hostilities ceased had been a preoccupation with Mackenzie King throughout the war. Canada must not recoil from Europe as she had done after the First World War. Abhorrence of war, isolationism, and neglect of armaments had not paid off between the wars. The weakness that these had caused had merely encouraged aggression on the part of others who had built up their strength. The best hope of preventing war lay in collective security; a combination of nations who felt the same would lead to strength: "The strong man armed," McNaughton was fond of saying, "keepeth the peace." Canada should be one of a large company of free nations, but her sovereignty must be respected; she must follow her own policies, co-operating alike with Britain and the United States. Canada, the Prime Minister said, must be "a nation wholly on her own *vis-a-vis* both Britain and the United States . . . we can never expect to have any recognition . . . in any other way." This, however, would be a difficult course to chart, for there were dangers to Canadian sovereignty from both these nations.

The most important operation in which I was involved during my two-year term as VCGS was the raising, organization, training, and dispatch of what was known as the Canadian Special Service Force (CASF). This consisted of an infantry brigade, with supporting arms and services, and became part of the British Commonwealth Division in Korea. During World War II, Japan had occupied Korea, and under the terms of surrender in 1945, Korea again became an independent but very disturbed and disorganized state. The country was eventually divided into two states, the Democratic People's Republic of Korea, Communist and Russian-dominated, and the Republic of Korea, supported by the United States. The dividing line between these two Korean countries was the 38th Parallel. For the next five years there was continuing friction between the two states, with North Korea being the aggressor. Finally, on June 24, 1950, North Korean forces invaded South Korea. The United States immediately called for a meeting of the Security Council of the United Nations, which was held the next day. At this meeting the Security Council approved a resolution calling for "the immediate cessation of hostilities and for North Korea to withdraw its forces from South Korea, and for all members of the United Nations to render every assistance in the execution of this resolution."

The North Koreans captured Seoul, the South Korean capital, and established a bridgehead over the River Han, about sixty miles south of the 38th Parallel. The United States acted without delay. U.S. President Harry Truman, on June 27, the day Seoul fell, ordered naval and air forces to the aid of South Korea. On the same day, the Security Council passed a resolution much more clear and definite than the previous one. This latter resolution said, in part, that since North Korea had not ceased hostilities, the Security Council "recommends that the members of the United Nations furnish such assistance to South Korea as may be necessary to repel the armed attack and to restore international peace in the area." Parliament in Ottawa was still in session, and in response to the above resolution, Lester B. Pearson, then Minister for External Affairs, reported to Parliament that Canada was conferring with other nations as to what part we should or could take in this Far East conflict.

On June 29, American ground forces were dispatched to assist South Korea. Truman was able to take this quick and positive action because he had forces immediately available.

On June 30, Parliament was to prorogue for the summer, and Prime Minister Louis St. Laurent made a statement that was of significance. He said, "Any participation by Canada in carrying out the United Nations' Resolution would not be participation in war. It would be our part in collective police action under the control and authority of the U.N. for the purpose of restoring peace. . . . It is only in such circumstances that this country would be involved in action of this kind." The position of the government was that, if the UN would say that such a police action under a UN command would achieve peace, Canada would consider a contribution. As a sort of assurance that Canada meant what it said, three destroyers would sail at once to the Far East to be available, if required. After this statement was made, Parliament rose for the summer. Three destroyers, *Cayuga, Athabascan,* and *Sioux,* sailed for Pearl Harbor on July 5, and on July 21 the cabinet approved the dispatch of an RCAF transport squadron of six aircraft to assist in the support of the UN force then in South Korea, which was composed entirely of American troops.

There was much discussion but little positive action by UN members, other than the U.S., to provide the required assistance. The problem was that Canada, like its wartime allies, had demobilized its forces to the point where the active Army was to have an establishment of only 27,000. When the Korean crisis had developed, the Army strength of regulars was just over 20,000, all ranks. The Army Council – CGS, VCGS, AG, QMG – and their staffs spent a great deal of time discussing plans of action in the event that the government decided to contribute ground troops to the UN Korean effort. This proved to be time well spent. On July 14, the Secretary General of the UN asked Canada directly if we could provide forces for service in Korea. Brooke Claxton immediately called a meeting of the Chiefs of Staff for their views on this request. The Navy considered that the three destroyers that had been dispatched to the Far East was a fair contribution. The RCAF offered an additional transport squadron of five transport planes. The CGS, speaking for the Army and voicing the views expressed at a meeting of the Army Council, favoured the contribution of an infantry brigade to form part of the British Commonwealth Division. The Minister took these views to a cabinet meeting on July 19. After this meeting, Prime Minister St. Laurent, in replying to questions by reporters, said that the dispatch of ground forces must have the approval of Parliament, and of course Parliament would not reconvene until early autumn! However, cabinet did approve the recruiting of personnel to fill existing establishments. The Army was about 4,000 under its authorized strength of 27,000. Consequently, on August 7, the Prime Minister announced the decision to recruit the Canadian Army Special Force. "It was to be specially trained and equipped to be available for use in carrying out Canada's obligations under the United Nations charter and the North Atlantic pact." The authorized strength was to be approximately 4,960, all ranks, plus a reinforcement pool of 2,105.

It was with regard to these figures that General Foulkes and I had a disagreement. He had already indicated to the Minister that the number of men required for a brigade would be 5,000. Whether from ignorance, which I doubt, or to encourage the government to make a commitment to provide a brigade, he had omitted to include the number for a reinforcement pool, which, of course, was absolutely essential. Foulkes was away from Ottawa, and I was in his chair when the Prime Minister wanted to make the above announcement. After consulting with and getting the agreement of the Adjutant General, it was I who gave the Prime Minister the number of men who would need to be enlisted. When Foulkes returned, he was livid with anger and seemed to think I had double-crossed him. In the evening he phoned me to come to his house. I did. He gave me a terrific blast for giving the Prime Minister the figures I had. I did not argue. But the first thing next morning I went to his office and told him I resented his comments and would be happy to retire. He said, "Forget it, Howard. I was tired last night." And that was the end of the fracas.

About this time I had been out to Western Command to witness an exercise being conducted by Brigadier John Rockingham, a Militia officer who had proven himself a good battalion and brigade commander during World War II. I had known him then and was impressed now with his conduct of the present exercise. Since most if not all recruits for the British Commonwealth Brigade would come from the Militia, I suggested to the CGS that the appointment of "Rocky," a Militia officer, to command the brigade would be justifiable and popular. And so it proved to be.

By August 18, about 7,000 had been recruited. Men had enlisted in great numbers as soon as it was announced on August 7 that the Special Force was authorized. On August 22, the problems we had had in the hasty enlistment of the Special Force were compounded by a strike of the Canadian National and Canadian Pacific Railways. This interrupted the dispatch of recruits to the various camps and to existing active force units for training. The railway strike, rather than the need to approve the formation of a Special Force and its dispatch overseas, made the recall of Parliament necessary, and this occurred on August 29. Finally, on September 9, the Commons passed the necessary Canada Forces Act, which authorized dispatch of the Special Force to Korea. There followed a further period of delay and confusion. The Americans under General Douglas MacArthur had driven the North Koreans back across the 38th Parallel and beyond. In the last week of October, General Foulkes was told by the U.S. Chiefs of Staff that, in view of the above, Canada need send only a token force of one battalion to Korea. The balance of the CASF, for the time being, could remain in training at Fort Lewis in the state of Washington. Twenty-two special trains carried the 25th Brigade, as it was now called, from various camps across Canada to Fort Lewis.

In accordance with the above recommendations of the U.S. Chiefs of Staff, one battalion, 2nd Princess Patricia's Canadian Light Infantry, was warned to sail for Korea on November 25, 1950, equipped with U.S. weapons and equipment. A few days before their departure, it was drawn to my attention that they would be giving up their trusted Lee-Enfield rifles and bren guns in exchange for unfamiliar American Garands and Browning automatics almost as they walked up the gangplank. As an infantryman, I was horrified. No one could be certain of the situation that the battalion would face when it arrived in the theatre of war. The last thing I was prepared to permit was their facing the enemy for the first time with weapons in which they had had no previous training. I stopped the exchange. As a result, the Canadian Army retained most of its own pattern weapons for the duration of the Korean War. It was a fortunate decision, for less than a month after the Patricias landed at Pusan they were in action against guerillas and had suffered their first casualty.

But the Korean conflict was not over, as General MacArthur and others had assumed. The intervention of Chinese Communist forces in aid of their North Korean comrades prolonged the hostilities for a further two years until July 27, 1953. The balance of our 25th Brigade in Fort Lewis was dispatched to Korea in three ships on April 19-21, 1951, and landed in Korea on May 4 to become a part of the British Commonwealth Division. An account of the operations of the 25th Brigade in Korea, from this time until the Armistice in July, 1953, is covered in detail in the excellent *Official History of Canadians in Korea* by Lieutenant-Colonel H.J. Wood. My direct concern with the organization and activities of the 25th Brigade and other duties as Vice Chief ended at the end of 1950. At that time I was appointed General Officer commanding Central Command with headquarters in Oakville, Ontario.

This move took place in January, 1951, and I was delighted. The two-year term as Vice Chief had been interesting and important, but I always felt remote and out of touch with the troops. Geographically, Central Command extended from the head of the Great Lakes to the Quebec border and from the U.S. border to James Bay. This area was divided into three military districts with headquarters in London for western Ontario, in Toronto (later moved to Oakville) for central Ontario, and in Kingston for eastern Ontario. Each district was responsible for training and administering the Regular and Militia units within their boundaries.

In addition to these districts, which came under my command, there were a number of corps schools located in Camp Borden, sixty miles north of Toronto, and at Barriefield, a suburb of Kingston. When I had been DCGS, the training programs for these schools were part of my responsibility. Hence I was quite familiar with their organization and knew most of the senior officers who commanded them. Now I was responsible for their administration. This was a major task because they included the following Corps establishments: Armour, Infantry, Army Service Corps, and Provost, at Camp Borden; and Signals – Electrical and Mechanical Engineers – at Barriefield; and the Canadian Army Staff College and Defence College at Tete de Point Barracks in Kingston. Camp Petawawa on the Ottawa River was another large establishment where one or more Permanent Force units were always stationed. This camp also was a centre for specialists' courses and summer training for both Permanent Force and Militia units. I followed the practice of my predecessor, Major-General Vokes, and spent the summer at Petawawa when the field training was at its height.

I spent four and a half years at Central Command, and I had few idle hours. Most of my time was spent away from my Oakville HQ, where I had an excellent staff who carried on during my absence on visits to all these camps, schools, and both Regular and Militia units. I was a great proponent of cadet training and attended many annual school inspections. I always spent some time at the large cadet concentrations in the summer at Camp Ipperwash on Lake Huron and at Camp Borden. In addition to some military training, we had a program of trade training in mechanics and electronics. But it was not all work, as the following clipping from a local paper indicates.

There Is a Much Much Better Way

James McNiven McSporran, nineteen, Scottish seaman, who, fed up with doing twenty-four hours continuous watchkeeping in hot weather, socked Vice-Admiral Eaton during an inspection, has been sentenced to two years in jail. It may be quite coincidental and possibly arranged long before, but the admiral has been promoted to commander in chief, West Indies Station.

At Ipperwash when the mercury rose to almost 100 degrees, a group of army cadets were due for formal inspection by Major-General H.D. Graham, DSO, who becomes chief of general staff in September. The general canceled the formal inspection and later went swimming with the boys in the lake. Everyone was happy.

With ordinary seamen or with cadets, in time of peace, it is best that everyone should be happy, or as happy as possible. We like the Ipperwash story because it is better to swim with a general than to sock an admiral.

Shortly after I left Ottawa to be GOC Central Command, there were some important changes in the hierarchy at Defence HQ. Lieutenant-General Foulkes was promoted to full general's rank and appointed chairman of the newly formed Chiefs of Staff Committee, and Lieutenant-General Simonds was repatriated from London where he had been on loan to the British Defence College and appointed Chief of the Army General Staff. So once again Guy Simonds was my boss. I am happy to say that, as had been the case in Sicily and Italy after Monty's intervention, we got along famously. He seemed to be satisfied with the way I ran my show at Central Command and, in fact, handed me a few tasks that normally would have been done by officers from his headquarters staff.

One of these was to chair a small committee to scrutinize the very large amount of paper being generated at Defence HQ and passed out to commands and schools of instruction, and from there, in whole or in part, passed on to subordinate units. I dubbed my task Exercise Paper Chase. With the help of two brigadiers, Jean Allard from Quebec and Frank Fleury from Ottawa, we did a thorough review. Believing that "a picture is worth a thousand words," we assembled on several tables in the Toronto Armoury copies of all the printed material emanating from headquarters in Ottawa. It was an astonishing display, much of it repetitive and quite unnecessary. We then made our report, recommending reductions that would save thousands of pounds of paper and countless man hours of time in writing, producing, and disseminating this unnecessary material. The CGS was pleased with the result of our labours and the flow of "bumpf" was greatly reduced. Another time he sent me, as chief Canadian representative, to Washington to attend a United Kingdom-United States-Canada discussion on standardization of small arms weapons and ammunition. One of my advisers on this occasion, strange to say, was Dr. Charles Best, co-discoverer with Dr. Frederick Banting of insulin for treatment of diabetes. Dr. Best was with me to give expert opinion on the effect on the human body of certain types of ammunition.

Again, Simonds gave me and my staff the task of planning and carrying through a divisional exercise in the Camp Borden area. This I called Exercise Red Patch. My principal general staff officer was Lieutenant-Colonel (later Major-General) Bruce Macdonald, and my chief of staff was Colonel (later Brigadier) George Wattsford. I mention them particularly because the exercise was a success, principally as a result of their skill and handiwork in preparing and conducting it under my overall supervision.

In January, 1954, Guy Simonds had been Chief of the General Staff for three years. I do not think they had been happy years. It must have been anathema for him to sit on the Chiefs of Staff Committee with Foulkes as chairman and senior in rank – a general as opposed to Simonds, still a lieutenant-general, although he had held that rank for almost a year before Foulkes. Foulkes had the ear of the Minister, unlike Simonds, who found it difficult to work with his political masters. I had the impression that Ralph Campney, the Defence Minister, actually disliked his CGS. Simonds, in spite of spending all his adult years in Canada and being a graduate of the Royal Military College, had been born in England and retained a very pronounced English accent. This in itself was a picayune matter and never bothered me, but to some people it was irritating. In addition to the accent, Simonds in many ways was more English than the English. Guy's talents were known and greatly appreciated by most of the senior British officers who, I would judge, looked upon him as one of themselves. Monty went so far as to say, "The Canadian Army produced only one General fit to hold high command in the Second World War – Guy Simonds." I think this statement was nonsense, but it probably reflected the opinion of many of Guy's English contemporaries.

I was astonished to receive a handwritten personal letter from Simonds, saying he thought he had been long enough in the Chief's chair and should retire. He was still about five years under the age limit of fifty-five for the retirement of generals. He also said in the letter that he had no other employment in view but thought he could find something. He then went on to tell me of some proposed appointments he intended to make at Army HQ and asked for my comments because he "had in mind" that I should succeed him. He closed his letter by asking me to treat its contents as confidential and, after reading, to destroy it. This I did, but I kept a copy of my reply, and this in part is what I said:

18 June '54

Dear Guy:

I have your personal letter of 13 June, and thank you for approval of leave and visit to I.C.I.B. [1st Canadian Infantry Brigade in Europe]. The second part of your letter was a distinct surprise.

I always assumed that you would stay on as CGS (or Chairman COS Committee) until you had reached retirement age. For my part, I had intended to seek retirement with effect 30th September, 1956 and never had any thought of promotion or appointment to CGS. At that time I will be 58. I do believe that there would be much grumbling, and justified, if I went up to CGS at age 57 or 58 and stayed there until age 60 or 61. On the other hand I am sure there will be no complaining if you stay until age 55.

Please do not think me ungrateful. I am deeply moved and greatly appreciate your confidence in me – but I cannot believe that the Army should lose your services in return for my promotion. . . . I have destroyed your letter as you requested.

With Kindest Regards,
H.D.G.

Here this matter rested for almost a year. In the meantime, my duodenal devil had not been idle. In early 1940 it had put me in hospital at Aldershot. In early 1942 it sent me to No. 15 Canadian General Hospital at Bramshot. In early 1944 it directed me to No. 1 Canadian General Hospital at Bari, and thence, via No. 10 in England, to Canada. In 1946 another attack sent me to the Queen Mary Hospital in Montreal, and in late 1950 still another to the military hospital at Rockcliffe. Why I was retained in the service during these repeated periodic attacks I will never know. Finally, in early October, 1954, I had been out on the divisional exercise "Red Patch" near Camp Borden for several days and sleeping in a cold caravan. I returned to my residence in Oakville for a weekend to help Jean entertain the Chief of the British Army Chaplain Services, who was on a tour. Our guest departed on Sunday afternoon, and that evening at dinner I suddenly had to rush from the table to the bathroom and vomit great quantities of blood. Jean, who was a nurse, got me to the nearby bed and telephoned Colonel Shier, the command medical officer, who lived only a few blocks away. He arrived within minutes, but in the interim, I had another violent hemorrhage and passed out. Within a few days I had had a gastrectomy, losing two-thirds of my stomach. At long last, after these almost fifteen years of intermittent suffering, I was rid of the devil and have not had a bellyache since! The cost is that I must avoid rich food. By year's end, I was declared fit by the doctors, who had done such an excellent job on my innards that I was soon back on my appointed round of duties.

In mid-February, 1955, I was in Ottawa and met the Minister privately at his request. He told me then that Simonds had told him he thought he had been long enough in the Chief's chair and should retire, as he was blocking promotion for other officers. Campney asked me if I would accept the position. I never did learn if Simonds suggested my name as his successor. I told Campney I would like to think it over for a few days and asked him if he minded if I discussed it with Simonds. He said he had no objection and requested that I write to him at his house address as soon as I had come to a decision. That same day I had a talk with Simonds in his office at Army HQ, and he repeated pretty much what he had told me in his letter written about a year earlier.

Back in Oakville at Central Command HQ, I pondered the matter for a few days. Unlike Guy Simonds, who repeated that he had no position in view and did not know what he would do on retirement, I already had a couple of legal-oriented jobs offered to me in the Toronto area. Further-more, the prospect of moving again to Ottawa from Oakville for a span of two or three years held little attraction for me, even though it meant a promotion to the rank of lieutenant-general. Yet the temptation to culminate my military service as Chief of the General Staff was hard to resist. Jean, ever co-operative, said she would be happy in whichever choice I made. So it was in this frame of mind that I wrote to the Minister on February 21, 1955, as follows:

My dear Mr. Campney:
Reference our conversation of the 15th, last Tuesday. It has been a problem to know what to do. My real difficulty is that I believe that Guy Simonds is probably a better C.G.S. than I would ever be and it seems wrong that he should retire whilst younger than I.

After leaving you I saw him for a few minutes and told him that you had suggested that on his retirement I should go up – no dates were mentioned. He said that he did not want to retire but felt that he should in order to avoid blocking promotion – apparently he has no other position in mind.

Well, the truth of the matter is that I would like the job for a two-year term but honestly hope Simonds will remain.

Perhaps I am over long in Central Command (4 1/2 years) and would be happy to retire to make way for the promotion of another officer. I am most grateful for your expression of confidence in me.
Sincerely yours,
Howard Graham

I mailed this to his house as he had requested. A week later I met Mr. Campney in London, Ontario, at a Royal Canadian Regiment function, and in a private talk he seemed a bit miffed (I could not blame him) and said he did not know from my letter whether or not I would accept the appointment. I said again I thought Simonds should stay. In answer to this he was very short. "Simonds must go," he said.

"Well, if that's the case, I will accept," I replied.

Talk about a reluctant bridegroom! In July, 1955, Simonds's retirement and my appointment were announced effective September 1, 1955. My métier was working with troops and, though I was somewhat remote from them as GOC of the large Central Command, I did have frequent contact with and much responsibility for both Militia and Regular units and the many school cadet corps that were then functioning in the area of my command. I knew from experience that in Ottawa, though I might be the top dog in the Army, I would have little opportunity to travel across the country to see the troops at work. So it was with a sad heart that I paid farewell visits to the many district camps, schools, and units that had been my responsibility for four and a half years. We left the excellent Army residence quarters in Oakville and bought a house in Rockcliffe Park, just behind Government House, where we would remain for the next four years.

13

CHIEF OF THE GENERAL STAFF

When I assumed the appointment of Chief of the General Staff in September, 1955, the government defence policy was and continued to be during my three-year term of office directed to three areas. These were concerned with the United Nations peacekeeping activities, the North Atlantic Treaty Organization (NATO), and the air defence of North America.

Canada was firmly committed to support the United Nations in peacekeeping efforts and to assist in these efforts where our government thought any proposed plan would prove to be useful. Our contribution would be within our competence as to manpower and cost. Under this heading, in September, 1955, we had commitments with the United Nations Truce Commission in Indo-China of 120 officers; a small detachment of seven officers with the United Nations Military Observer Group in Kashmir; and four officers with a United Nations Truce Commission in Palestine. All these personnel were rotated after one year's service. In addition, we still had in Korea and Japan about 300 personnel, the remnant of the Brigade Group that we had provided as a part of the British Commonwealth Division during the Korean conflict. This small group was brought home in mid-1956, and our United Nations commitment in that area, which had begun during my turn as Vice Chief in 1950-51, came to an end. Ahead of us was involvement in another United Nations peacekeeping effort in the Middle East.

Canada was also committed to maintaining support to NATO. This involved the provision of an infantry division of three brigades with supporting arms and services and the necessary headquarter organizations. One brigade under this commitment had been organized, equipped, and dispatched to Europe in 1952. The balance of the division was to be organized, trained, and ready for action, but was to remain in Canada until called for by the NATO Supreme Commander in Europe (Saceur). It would then be dispatched as soon as shipping was available. This plan was the principal reason for developing the large permanent Army base at Gagetown in New Brunswick, so that the troops would be near an eastern Canadian port. The personnel in the brigade stationed in Europe during peacetime took with them their wives and families. The policy was for them to remain overseas for a three-year period and then return to Canada and be replaced by a second brigade with their dependents. This exchange was a major and expensive operation. The total number of persons moved from October, 1955, to January, 1956 (the period of the first changeover), was just under 20,000 and involved the charter of eleven ships. This system was altered during my period as CGS so that only one-third of the troops and their dependents were moved each year.

Canada was committed, finally, to a policy of North American defence, and this largely fell to the Air Force. The Army commitment was for a mobile-striking force with a strength of about 3,500, composed of three parachute battalions with 4.2 mortars, engineers, signals, and supporting units. We were at this time, and probably are today, dependent on our neighbour to the south to come to our aid in the event of a hostile encroachment or attack upon Canadian territory.

It will be appreciated that the Army commitments, under our three-pronged defence policy, required static establishments such as headquarters, schools, hospitals, depots, repair shops, and other administrative units. In addition to these manpower requirements, the Regular Army, and specifically the Royal Canadian Engineers, still had the responsibility for supervising the maintenance of the North West Highway from Grand Prairie to the Alaska border. The total strength of the Regular Force was about 50,000 from September, 1955, to September, 1956. The government's defence policy can be said to have changed only by adding to Army responsibilities the task of aiding the Civil Defence organization, should that be found necessary. The Army's commitments under the above defence policy changed somewhat in two regards. The Korean force (the UNSF) was wound up, and the personnel were either discharged or re-enlisted in the Regular Army. The government in 1956 agreed to provide a force of up to 1,500 to a newly organized United Nations Emergency Force (UNEF) in the Middle East. This is the way the emergency arose and the way we became involved.

The Egyptian president, Colonel Nasser, on July 26, announced that Egypt had nationalized the Suez Canal and would use the proceeds from the tolls to help finance the building of the Aswan Dam. This was a prime project Nasser considered would result in great advantage to the economy of his country. As one would expect, Nasser's action aroused, particularly in Britain and France, feelings of intense anger. Anthony Eden, the British Prime Minister, intimated that if necessary his government would take military action to restore control of the canal to the former owners or to some other international group. An extremely dangerous confrontation was shaping up between the United States, Britain, and France, on the one hand, and the Arab states, Egypt, and the Soviet Union, on the other.

At this stage, toward the end of September, Britain and France asked for a meeting of the United Nations Security Council to consider the situation, and Egypt did likewise. In the next month the Security Council sought to find a solution that would be acceptable to all parties concerned but met with no success. Then Israel entered the fray on October 29, 1956. The state of Israel had been carved out of Palestine and was the subject of raids over its borders and antagonistic policies and actions by the Palestinians and their friends, including Egypt and other Arab states. Israel responded to these breaches of the peace by retaliation on a basis of "ten for one." The Israelis evidently thought this was an opportune time to take the ball far into the Egyptian court. They made a surprise attack which, within days, carried their forces across the Sinai Desert to the shores of the canal. Britain and France immediately issued an ultimatum that unless both Egyptian and Israeli forces ceased hostilities and withdrew ten miles from the canal within twelve hours, they would move in and occupy key positions to keep canal traffic moving. The ultimatum was rejected by both Israel and Egypt, and on October 31 the French and British air forces began to bomb targets in the canal zone. By November 1, an emergency session of the UN General Assembly was called to deal with the grave crisis. Facts that became known later indicated that France and Britain urged Israel to take the action it did against Egypt.

Our Minister for External Affairs, Lester B. Pearson, was sensitive to the possibility that the Suez affair could become a conflagration that would spread far beyond the Middle East. It was Mike who took the initiative and proposed, with the full support of his cabinet colleagues, to the UN General Assembly on November 2 that a United Nations Emergency Force be formed to stand between the Israelis and the Egyptians and that the British and the French forces withdraw and thus give time to find a political solution that would be acceptable to all parties. On November 5, Egypt formally accepted Mike's resolution, which had been agreed on by the Assembly, providing for a UN force. On the same day, Israel unconditionally accepted the ceasefire that the Assembly had demanded. The next day, the Secretary General, in discussing the formation of the UN force with Mike, estimated a strength of 10,000 would be required and suggested that Major-General E.L.M. Burns, a Canadian, should be the commander of the force. The Canadian cabinet, on the recommendation of Ralph Campney, our Defence Minister, agreed to a Canadian contribution of a battalion group of 1,000 to 1,500 men, and Mike hoped that they would be available within ten days or two weeks.

Major-General E.L.M. Burns was appointed Chief of the UN Command about November 4. He had been in charge of a UN Palestine Truce Supervisory Organization along the Israeli-Palestinian border and was therefore familiar with the area and the nations involved. The choice of Major-General Burns, "Tommy" as he was called, could not have been bettered. He was perfectly bilingual in French and English, had held high command during World War II, and had experience in both civil and military affairs to meet and discuss problems with Nasser, Ben-Gurion (the president of Israel), the Secretary General of the UN, and others in high places.

During all of this the Canadian chiefs of staff were kept informed through the Department of External Affairs as to how Mike was getting on with his initiative and what might be expected from Canada as a contribution to the peacekeeping force. I recall sitting with Bud Drury, the deputy minister of defence, in his office drafting a letter for our Minister saying what Canada was prepared to provide. Bud, as a lawyer, was a stickler on the fine differences in words. I became very impatient after half an hour of indecision on his part as to whether we should use the words "we offer" or "we proffer"! Was there a difference? I did not think so; either would do. He was not sure. I do not know what he eventually decided, as I walked out. It was well past dinnertime!

Early in November, Air Marshal Slemon, the Chief of the Air Staff, and I flew to New York in an RCAF Dakota and spent a day with Mike and Ralph Bunche, the deputy secretary general, at United Nations headquarters, discussing the possible composition and size of the force that might be needed. At the same time, our Chiefs of Staff Committee, knowing that the Canadian portion of the UN force would need quickly available sea transport as well as air transport to move men, equipment, and stores to Egypt, asked the Chief of the Naval Staff, Vice-Admiral Harry DeWolf, to order the Canadian aircraft carrier HMCS *Magnificent*, which was on a naval exercise in European waters, to return to Halifax at once. She was berthed in Halifax within six days of our request. A few days later Slemon and I again flew to New York to see Mike and to talk to Major-General Burns, who had returned from the Middle East the day before. On this trip, we took the Army Director of Staff Duties, Colonel George Leach (later Brigadier), who would be of great assistance to Burns in preparing in detail an establishment of the troops required and the organization of an administrative group that would be responsible for the provision of medical, signal, catering, repair, and all the services for a varied international force. On this occasion I recall sitting in an anteroom of the General Assembly, which was in session, with Pearson, Slemon, and Leach, and was surprised to see Pearson take a cigarette from a packet and rather clumsily light it. "I didn't know you smoked, Mike," I said.

"I really don't, but I'll be taking dope if we don't soon get this show on the road," he replied, shaking his head as if in desperation.

The outcome of the meeting with Burns and Pearson was that an ad-hoc establishment was worked out for the whole multinational force, i.e., the strength, types of equipment, and supporting services needed. It was proposed and agreed that the Canadian contribution would be one infantry battalion. We lost no time in flying back to Ottawa and reporting to our Minister, who would put the plan to the Prime Minister and cabinet for approval. I then telephoned General Vokes in Victoria and told him that no decision had yet been made by the government, but in strict confidence I said that we might be moving the 1st Battalion of the Queen's Own Rifles in a hurry. It would be helpful if he could have a thorough medical check of the personnel done as a routine matter because, since no firm decision had been made, we did not want a lot of conjecture by the media.

A day or two later cabinet agreed to the proposal. It was a Thursday evening, and I again contacted Vokes and told him the QOR were to be airlifted by the RCAF to Halifax with light equipment and transport. Tri-service plans were worked out in the Chiefs of Staff Committee for the air movement to Halifax and joint naval-air movement of the battalion and its light vehicles to Egypt. Orders were issued for the movement of trucks from Hagersville, Ontario, where there was a large stock of vehicles carried over from wartime, by train to Halifax. By Tuesday morning, the battalion of the Queen's Own Rifles, about 800 strong, had been airlifted and were in Halifax ready for embarkation. This in itself was no mean feat when one considers the distance involved and the work required, all accomplished within seventy-two hours.

Then the axe fell. Colonel Nasser had been reluctant to have a UN force in Egypt and insisted on approving the nations that would form it. None of the major powers was to be involved. His hatred of Britain was intense, and when he learned that the proposed Canadian part of the force was to be a battalion called the Queen's Own Rifles, he refused to have them in his country because of the British connotation. We were in a quandary because other available units, like the Princess Patricias, the Royal Canadian Regiment, the Black Watch (Royal Highlanders), and the Royal 22nd Regiment, would all have had the same effect on the testy Colonel because of the implication that they were British. The difficulty was resolved by changing our commitment from an infantry battalion to the provision of the supporting and administrative services. This involved the selection of such specialists as signallers, medical and dental personnel, motor mechanics, clerks, cooks, and others from various establishments across Canada. This was a much more time-consuming task than simply ordering the move of one unit. But it was expeditiously done. The staff work and co-operation of the Air Force were superb, and in a few days the greatly disappointed QORs were returned to Calgary and the administrative force was dispatched from Halifax, some in HMCS *Magnificent* and an advance party in twin-engined C119 RCAF aircraft, which made the long flight by refuelling in the Azores.

Our contribution proved of great value, probably more so than an infantry battalion would have been. As a matter of fact, Major-General Burns later wrote to Mike to say, "As things have worked out, the difficulty about sending the Queen's Own Rifles was a blessing in disguise, since the Canadians in the base units have made all the difference in the world in the efficient operation of the administration side of the Military effort. We just could not have done without them."

Naturally I was gratified to hear this. But the greatest credit must go to Mike, who, by dint of his initiative, determination, dedication, and weeks of persistent negotiation, brought an end to the confrontation between Egypt and Israel and the possibility of involvement by the superpowers. His work was justifiably rewarded with the Nobel Prize for Peace. It also gave me much pleasure to be able to send a personal message to Tommy Burns just before Christmas, 1957: "Minister is announcing today your promotion to Lieut.-Gen. effective 1st January, 1958. Well deserved – my congratulations." The force he commanded with success and distinction came from New Zealand, Pakistan, Poland, Colombia, Sweden, and Norway as well as from Canada. To succeed as he did in restoring peace and stability, at least for a short period, in that troubled area, surely deserved recognition and reward by promotion to that rank.

The organization of the UNEF was an important event during my term of office as CGS and took up a great part of my time and thought. But there were other matters of equal or perhaps greater importance that required attention. Our commitment to NATO in 1955 was an infantry division with all supporting arms and services, of which one brigade group was in Europe and the balance of the division was to be trained and ready in Canada to be moved by sea as soon as shipping was available. The divisional organization was to be similar to that used in World War II ten years earlier.

The development of missiles was to have a profound effect on our concept of both strategy and tactics and the types of organization and weaponry that would be required to meet the missile threat. First there was the short-range missile that had been used by the Germans near the end of the war. I well remember the consternation with which we learned of the Russian long-range weapon, soon to be dubbed the ICBM, the intercontinental ballistic missile. It was followed in short order in both the U.S.S.R. and the U.S.A. by increased accuracy of missiles over long ranges by a guidance system, by missiles with ever-increasing destructive power in atomic warheads, by new methods of launching from surface ships, from submarines, and from aircraft. Coincidental with the development of the offensive weapon, we saw the production of defensive missiles that would intercept and destroy the attacker, or so it was hoped. The concept of meeting a missile assault on the fields of Europe with infantry, armour, and artillery equipped and organized as in World War II lost its attraction. Smaller, more flexible, highly mobile formations for ground forces seemed to be the answer, with an air component of low-flying, short takeoff and landing aircraft to aid in supply and movement.

In line with the above concept, we reorganized the Regular Army field force by forming four brigade groups requiring in total twelve battalions of infantry. (We had fifteen but disbanded three.) With the manpower thus saved, we formed another armoured regiment, First Eighths (1/8) Hussars of New Brunswick, to be added to the Royal Canadian Dragoons and the Lord Strathcona House. We were still one armoured regiment short. To give the 4th Brigade Group an armoured element, we would transfer one squadron from each of the three regiments to the 4th Brigade Group. Artillery and other arms and services were to be the same in each brigade group, except that a parachute element would be maintained in the Western Brigade. These changes were considered under the Liberal government, when Ralph Campney was Minister of Defence, and completed under the Conservative government elected in the summer of 1957, when George Pearkes, VC, became Minister.

The changeover from a Liberal to a Conservative government in mid-1957 resulted in no alterations to our defence policy or our commitments under that policy. However, it did involve a considerable amount of extra work in briefing the new Minister on details of Army programs and organization and travelling with him at various times to visit Army units and establishments across Canada. I had known George Pearkes for many years, of course, and served under him when I was commanding my regiment in Britain when he commanded the 1st Division. Unquestionably he had been an efficient and gallant soldier and was a most pleasant and friendly Minister, but I found him less ready than Campney to make changes. Being an old professional soldier himself, and as such having exercised command over his present CGS, he probably considered that he knew better than I did what was best for the Army. And perhaps he did! But we were good friends, and he and his wife, Blytha, often came to our house for an evening of bridge.

A government venture in the field of aircraft development that was underwritten prior to and during my term as CGS was the CF105, better known as the Arrow. This supersonic fighter was a development project at A.V. Roe Canada Ltd. AVRO began it in 1953 under a contract with the Liberal government of Louis St. Laurent. An initial appropriation of federal funds was $30 million, but it was anticipated at that time that the program of development would cost $100 million and that the cost of each aircraft after this expenditure would be about $2 million. It is interesting to note that as early as 1953 C.D. Howe, the Minister of Defence Production in the St. Laurent cabinet, said that this program gave him the shudders! Had he foreseen the taxpayers' money that would be poured into the project during the next five years, he would have had more than the shudders! By 1959, $335 million would be spent, and AVRO was still not ready to go into production. It was estimated that when, and if, the AVRO Arrow was ever put in service in the relatively small Royal Canadian Air Force, the cost per machine would be in the range of $12 to $13 million, as against the original estimate of $2 million. While Minister of Defence in the St. Laurent government, Campney went to Washington expressly to seek a commitment from the U.S. Defence Department to buy our Arrow, which no doubt would have reduced the unit cost. However, the United States had its own aircraft industry to consider, and while they admitted that the Arrow looked as though it would be an excellent supersonic fighter, they were not prepared to make a commitment to buy it.

Air Marshal Curtis, who was the innovator and principal proponent of the Arrow program, retired in 1953 as Chief of the RCAF and immediately went to work for AVRO as vice-president. (I always questioned, in my mind, the propriety of this action.) Air Marshal Slemon, Air Force Chief from 1953 to 1957, and Air Marshal Campbell, who succeeded Slemon in 1957, as one might expect, continued to support the program, even at prohibitive costs that consumed the greater part of any reasonable defence budget. The Navy and the Army, through their chiefs – Admiral Maingay (followed by DeWolf) and Lieutenant-General Simonds (followed by Graham) – and the Chairman of the Chiefs of Staff, General Foulkes, argued against it, and so advised defence ministers (Campney and Pearkes).

The matter came to a head in August, 1958, when Prime Minister John G. Diefenbaker called a meeting of the Chiefs of Staff Committee and the Cabinet Defence Committee at the Parliament Buildings. I remember it well and always wondered why no reference was made to the historic meeting in all the books and articles that have been written for and against the decision to scrap the Arrow. The Prime Minister gave a brief and accurate history of the problem and said that before a decision was made on the fate of the project he wanted to give this opportunity to those present to express their views. Air Marshal Hugh Campbell was present but, being the realistic and intelligent person he is, surely realized that any argument he could put forward for the continuance of the project would bear little weight. In any event, he had argued strenuously in its favour with the Chiefs of Staff Committee but had found no support from the rest of us.

The cost of the development and production of the aircraft was one factor against it. Another was the advent of the missile as both an offensive and a defensive weapon. It was argued that the missile would make the use of a supersonic interceptor like the Arrow of less importance in the future. In my opinion, the cost factor was the decisive one in 1957. Finally, the Prime Minister announced in the House of Commons on February 20, 1959, "The conclusion reached is that the development of the Arrow aircraft and Iroquois engine should be terminated now." From that day to this, arguments flow back and forth as to the wisdom of the decision. The immediate discharge by AVRO of more than 14,000 workers, many of them highly skilled technicians and scientists, was a sad blow to them and to the community, and to the engineering and research activities in Canada. I have every sympathy for those who suffered, and perhaps still suffer, as a result of that decision. But I felt that the government's decision was courageous and correct.

The organization and function of the Regular Army occupied much of my time. But time was always found for the Militia, the citizen-soldiers of Canada. The Militia have a long and brilliant record of achievement. Originally organized by the earliest settlers in this country for the protection of their homes and lives, the members of the Militia during the eighteenth and nineteenth centuries on many occasions, either alone or in conjunction with British Regulars, did indeed drive out the invader. In 1914, and again in 1939, when our Regular forces were very few in number, it was upon the units of the Militia that we built up several divisions for service overseas and several more for home defence. Those who went overseas were, in fact, defending their homes and families just as their forebears had done a century or more before.

During my term as CGS, we had adopted a defence policy with commitments that required a full-time Regular Army strength of about

50,000 as opposed to the 4,000 we had prior to 1939. New weapons, principally missiles with atomic warheads of tremendous destructive power and intercontinental range, had been developed. This continent, for the first time in history, might be a target for direct and accurate attack. Our effort was then, as it has continued to be, the prevention of war. If, in spite of our efforts, we are drawn into a major conflict, one of our first and most vital objectives would be for survival within our own boundaries. The situation would certainly be quite different from the ones that existed in 1914 and in 1939, when we had many months and even years to complete the training of tens of thousands of Militia men and new recruits and dispatch them overseas.

In any future conflict, ground forces will be engaged in two types of battles. One will be direct contact with enemy ground forces in Europe. This battle must be fought by forces in being and *in situ* and will be of short duration. The other battle will be in this country to assist in restoring order out of the dreadful chaos and state of anarchy that may exist as the result of enemy attacks with long-range nuclear weapons. In this second important type of battle, in my opinion, the Militia and the Regular forces then existing in Canada, acting in conjunction with the Emergency Measures Organization (EMO), will be needed. The Militia must continue to receive some military training because it is the military-type of training with its built-in disciplines that makes Militia men so very valuable in giving aid to the civil authorities.

Many of the military skills are precisely those skills needed in time of great disasters – medical assistance and first aid, communications, traffic control, engineering, repair services, etc. But wars do not always happen as one expects, and there is another reason for continuing to give military training to members of the Militia. In the event of war, it will be of great value to have a reserve of at least partially trained personnel in all arms who may be more quickly brought up to a higher standard of training and be able to assist as reinforcements, in static training, or in administrative establishments. The period 1955-58 saw a greatly increased emphasis on training the Militia in civil defence duties and on establishing an association with EMO, as it came to be called. But the role was not quietly accepted by the Militia, and I had not expected that it would be. It was accepted by the government, however, and I did my best, with some success I think, in explaining to the Conference of Defence Associations (CDA) the reasons for this policy.

Another radical change in Militia affairs was the reduction of the annual number of days training from sixty to forty-five and the discontinuance of Militia camps. In lieu of attending their own camps, Militia units would now send suitable personnel to train each summer with Regular units for a specified number of days. In addition, courses of instruction would also be given to specialists, e.g., signals, mechanics, engineers, etc. On speaking to one of the Militia corps associations, I recall being asked why the period of training was cut by fifteen days, and why summer camps were being eliminated. I replied that we must accept the fact that the Canadian taxpayer could be expected to pay only a certain amount on defence, as funds were needed for the Regular forces to be properly equipped to meet our commitments, so by so doing we were, hopefully, helping to avert war. As to discontinuing Militia camps, I said the money saved was substantial and, furthermore, the training of Militia with Regular units would, in fact, result in better training than what they received in their separate camps. I asked them to think about conditions as they then were in 1957 and to then think about the type of war that we were likely to have to fight in the future, rather than to continue to think in terms of conditions and types of war in the 1930s and 1940s. I said I appreciated that it was difficult to break from long-established custom and tradition, but if we were going to prepare for the next war instead of the last one, we simply had to change our thinking and our plans to fit the new set of circumstances. I came away with the impression that most of those present were satisfied that the new role and training policy for the Militia were realistic and justified.

Of the many decisions I had to make as the CGS, one that gave me much difficulty was the selection of officers for promotion to senior ranks, i.e., brigadier and major-general. There might be a vacancy for one major-general, and five or more brigadiers would be shown in their annual reports to be suitable for promotion. The easiest action for me to take would be simply to recommend that the Minister appoint the most senior of those brigadiers. But from my personal knowledge, and after consultation with the AG, I might decide that a brigadier who was third or fourth down in seniority was the best man for the major-general's job. It can never be said that I was biased in my final decision by personal friendship or any other consideration except "what is best for the service." Quite often, as one might expect, an officer who had been passed over two or three times would request an interview with me or even, as a last resort, with the Minister. I always granted the interview and explained the reason for my decision, but I am afraid this was of no avail in satisfying the officer.

Perhaps I was a little old-fashioned, but when I went to Ottawa as Chief, I thought it proper that I should make a formal call on the Governor General, the Hon. Vincent Massey, one of whose titles was, after all, Commander-in-Chief of the Armed Forces. I telephoned his son Lionel, who was also his secretary, made an appointment, and in a day or two reported at Government House. I met His Excellency for the first time over tea. I think he appreciated this little act of courtesy and perhaps it had an influence on some of my future activities. As we had bought a house just across the road from the rear entrance of Government House, quite often on a Sunday afternoon Lionel would telephone and ask Jean and me over for supper with His Excellency and to see a film afterwards. Invariably it was a Danny Kaye picture, Kaye being His Excellency's favourite film star. Mr. Massey, I found, as one might expect, a most gracious, kindly, and learned gentleman. He was always keenly aware of the fact that he was the Queen's representative, and it seemed to me that sometimes he felt that some of the royal *éclat* had rubbed off on him! He expected the ladies to curtsy and the men to bow their heads when they were presented. After a formal dinner, as the ladies rose to retire, he stood near his place at the head of the table and the ladies, each in turn, curtsied before leaving the dining room. The men stayed awhile for an informal chat with His Excellency over their liqueurs and coffee. This was the custom of royalty, and Mr. Massey took the position that he was, indeed, standing in for Her Majesty. In recent years, as customs have changed, much of the dignity and prestige of this office has been lost.

Queen Elizabeth and Prince Philip came to Ottawa for the opening of the first Parliament of the Diefenbaker term on October 14, 1957. It was a colourful ceremony and, with the other Chiefs of Staff and Chairman, I was in the small group that escorted Her Majesty and the Prince to the Senate Chamber. In the evening a dinner was held at Government House for members of the cabinet, Supreme Court justices, *et al.* I was not invited, but early in the evening Lionel telephoned and asked Jean and me to come over after dinner to meet Her Majesty. We were glad to do so, and about 9:30 p.m. we walked across the road to Government House. I was in uniform and Jean, of course, in formal frock. An aide received us and asked us to stand at the farthest end of the ballroom where many other after-dinner guests were assembled. When the Queen and Prince

Philip came into the ballroom after dinner with Mr. Massey, they moved slowly through the gathering, shaking hands and saying a word to those whom Mr. Massey introduced. Finally, they came to the end of the room where Jean and I were standing. Mr. Massey introduced us, and the five of us had a pleasant chat, mostly about where I had served in both world wars and about our time in London in 1946-48. I learned later that the Queen had accepted an invitation from Prime Minister St. Laurent, before he was superseded by Mr. Diefenbaker, to come to Canada to officially open the St. Lawrence Seaway in 1958 or 1959. Mr. Massey would have known of this arrangement, and perhaps had in mind that I would be a suitable person to take charge of the arrangements. Perhaps he wanted the Queen and Prince Philip to get some idea of what kind of bloke I was!

Having completed the normal three-year term as CGS on September 1, 1958, and in my sixty-first year, I looked back with little nostalgia, but certainly with many memories, happy and otherwise, on almost forty years of Army service. I had served in every Army rank from private in World War I to lieutenant-general in the post-World War II era. I had held every rank except that of warrant officer – i.e., sergeant-major. (I was never quite good enough for that!) I had learned from my experiences as a private and NCO what those men expected from their officers. It was sympathy and understanding of their problems, strict discipline, reasonably and fairly applied, valour, and proficiency. These attributes are what breed high morale and first-class units. I would take it as a high compliment if a veteran of one of my units said of me, "He was a tough old s.o.b., but he never asked us to do anything he wouldn't do himself."

When I took over from Lieutenant-General Simonds, I found, as I expected I would, a first-class organization designed to carry out the commitments made by the government. Some changes had to be made during my term of office, but they were by reason of the development of new weapons and to conform with NATO plans and policies. But plans to send the balance of an infantry division to post-war Europe never seemed sound or realistic to me. My belief was that, with newly developed weapons and techniques, one side or the other would have to hoist the white flag of surrender within days. I agreed with Mr. Pearson's statement made when he accepted the Nobel Peace Prize that "the stark and inescapable fact is that today we cannot defend our society by war, since war is total destruction and if war is used as an instrument of policy, eventually we will have total war. Therefore the best defence of peace is not power, but the removal of the cause of war and international agreements which will put peace on a stronger foundation than the tower of destruction." But until such agreements were reached, I agreed with McNaughton's repeated axiom, "The strong man armed keepeth the peace."

On September 4, 1958, I received a letter from Ralph Campney, which gave me much pleasure after my three years as CGS: "On the occasion of your relinquishing the important position of Chief of the General Staff, I should like to express to you my personal appreciation of the excellent way in which you carried out these duties during the period when I was Minister of National Defence. The understanding, energy, and common sense with which you carried out the exacting duties of that position were a source of comfort and of strength to me as Minister. You have served Canada well in that, as well as in other capacities, and I think it opportune that I say so at this time."

I always considered that there should be a close relationship between the Civil Defence Organization and the military forces. When it was announced that I would retire as Chief of General Staff at the end of August, 1958, Waldo Monteith, the Minister of Health and Welfare, asked me if I would accept the appointment of Director of Civil Defence, for which he and his department (strange as it may seem) were responsible. At the same time, Brigadier James Melville, chairman of the Pension Commission, asked if I would fill a vacancy there, and the Department of Veterans' Affairs offered me a position with the possibility of becoming deputy minister. It was gratifying to have these opportunities offered to me, but Jean and I had decided that we would settle in the Toronto area, preferably Oakville, where we had relatives and many friends and I had offers for employment. Mr. Monteith then asked if I would, as a temporary task, review the whole civil defence structure and recommend such changes in its organization and functions as I might think desirable. This I agreed to do from September to the end of the year.

Having turned over my position as CGS to Lieutenant-General S.F. Clark the day before, I set off on September 1 for the east coast. During the next four months I visited more than forty cities across the country. In each province I was accompanied by the provincial civil defence co-ordinator and I met with the premier and the cabinet minister responsible for civil defence. There was no provincial organization in Quebec because the provincial government considered civil defence to be a federal responsibility. I did not visit the northern Territories but I ascertained through my survey all the relevant information and viewpoints. What I discovered was recorded in my report, dated December 31, 1958. I explained in the conclusions:

(a) That the U.S.S.R. had a capability to attack any target in North America with nuclear bombs delivered by aircraft;
(b) That by 1961 they would be able to attack with missiles launched from surface ships and submarines and would have in production missiles with a range of 3,000 miles;
(c) That such bombs or/and missiles would be able to penetrate our defence in numbers sufficient to devastate Canada;
(d) That, on the other hand, the Western powers had a capability equal to or greater than that of the Soviets, and it must be assumed that this situation would continue.

I went on to note the following:

The mere possession of a capability does not in itself constitute a threat. One must give some thought to the "intent" behind the capability. In the case of the Soviets, it is difficult, of course, to appreciate their intentions. However, it is generally conceded that the USSR has not the intent, as a deliberate act, to launch nuclear war upon North America, whilst the United States has the power, as she now has, to retaliate in kind. Such an attack by the Soviets and retaliation by the USA would result, it is believed, in the virtual destruction of both countries, and Canada.

It has been suggested that the USSR might have the intent to attack if she thought the retaliatory capacity of the United States could be destroyed before it became effective, or could be so badly damaged as to make it ineffective against the Soviet home defences. This suggestion may have merit, but the possibility of the retaliatory capacity of the United States being destroyed or being made ineffective is very remote, indeed.

Political and Military leaders in the Western block have repeatedly said that they do not think a nuclear attack will be made upon North America so

244

long as the USA retaliatory capability is maintained. *E.g.*, Field Marshal Montgomery has written, "We must understand that Russia is just as frightened of attack by the Western Alliance as we are of attack by her." She, therefore, has her deterrent . . . it is clear that the launching of war by Russia against the Western Alliance can be considered unlikely in any future that we can foresee, *provided we do not reduce the overall deterrent.*

My conclusion, therefore, was that an attack was unlikely, but if it came, it would follow a period of tension during which there could be population dispersal. But once hostilities commenced, the period of warning for any or all parts of the country would be a matter of minutes, or, at most, two or three hours.

In my interviews I found that in most cases the premiers and ministers responsible did not know much about their civil defence organization. They left it to the co-ordinator. There was no interprovincial co-ordination. I also interviewed Red Cross, St. John's Ambulance, and Legion officials, not to mention mayors and the director of civil defence of the United Kingdom, and came to the conclusion that the mass evacuation of large cities was not only impracticable but unacceptable to the population. I was dealing with a nuclear attack and not with a local or natural disaster like a flood, a fire, or an explosion. Plans for the removal of people from such an area were made by local authorities under the assistance of the Emergency Measures Organization (EMO). In my report I was dealing with the survival of the nation. I learned through military attachés that in the Soviet Union it was compulsory – and perhaps still is – for adults to take a number of lectures on civil defence. In the West the public was little interested in civil defence. I said little or nothing about a shelter program because I felt that this should be dealt with by a newly designed civil defence or survival organization.

I recommended that the federal government should assume sole responsibility for, and bear the cost of, civil defence, just as it does for the armed forces; and that the responsibility should be with the Department of National Defence. I further recommended that a Director-General of Civil Defence be rated at the level of a major-general. This recommendation was acted on. My report was dated December 31, 1958, but it was not delivered until mid-January, 1959. The reason for the delay has to do with "important people."

One afternoon late in October, I was meeting with officials in Leth-bridge, Alberta, discussing the problems of civil defence, when I was called to the telephone. It was General Pearkes calling from Ottawa. He said some important people were coming to Canada and Diefenbaker wanted me to take charge of the arrangements for their visit. I knew he was referring to the Queen and Prince Philip. I said I preferred not to get involved, but if the PM thought it essential that I do so, I would be willing. I checked into the Bessborough Hotel in Saskatoon for the last weekend of October. There I received a telegram saying that the PM would be at the same hotel, arriving Saturday evening and staying until Monday, and would like to see me. By chance we met in the elevator on the way down to dinner that evening, and he asked me to come to his suite at 10:00 a.m. the next morning.

This I did. His wife Olive was with him, and he was lying on a couch in his shirt sleeves with his feet elevated and a mass of papers on a coffee table beside him. "Good morning, General, this is the way I like to work," he said.

"Almost like Mr. Churchill," I replied, and this seemed to please him. (Mr. Churchill was fond of working as he lay in bed and received many visitors as he lay in a recumbent position.) For an hour or two we discussed the sketchy plans that had already been prepared for the Queen's forthcoming visit. The PM said the original purpose of Her Majesty's visit was to officiate at the formal opening of the St. Lawrence Seaway in midsummer. However, at the government's request, she and Prince Philip had agreed to visit all parts of Canada, including the Yukon and the Northwest Territories. Our ambassador to Holland, H.F. (Temp) Fever, had been brought home to do the planning, but he was needed in the Department of External Affairs, so there was no one to take on the task.

I reminded the PM that I still had much to do on the civil defence survey. "Oh," the PM said, "it won't matter if that is delayed a bit, and anyway, the Queen's visit shouldn't take much of your time until next year."

Little did he or I know the complexities and the extent of the task I had agreed to undertake. I cut short some of my scheduled meetings between Saskatoon and Ottawa, but fortunately I had enough material to prepare my report, which involved sixteen- to eighteen-hour days, including weekends.

On Thursday, October 30, I was back in Ottawa and went to see

Robert Bryce, the Clerk of the Privy Council, who had put together the sketchy preliminary plans the PM and I had discussed in Saskatoon. Bryce, whom I knew well, added more detail on what he had accomplished with the External Affairs people who had been consulting with the U.S. ambassador in Ottawa regarding the plans for the Seaway opening. Bear in mind that the Seaway was a joint Canada-United States project, and therefore both countries should agree on the site and program of the opening ceremonies. J.H. Cleveland was on the American desk in our Department of External Affairs and was of great assistance to me in arranging and attending our joint meetings, of which there were many.

Bryce told me that Prince Philip was in Ottawa attending a meeting of the English-Speaking Union and was staying at Government House. Bryce had arranged for me to see Prince Philip and Governor General Massey the next morning, Friday, October 31. This was an interesting meeting. Esmond Butler, a Canadian from Manitoba whom I met then for the first time, sat in on the meeting. Esmond had been on Massey's staff, but was now assistant press secretary to Her Majesty, and he was to be of immense help to me in the ensuing months. Prince Philip briefed me on a great many points for the detailed program of the lengthy trip across Canada. His advice was to make the tour as personal as possible. The Queen would like to see some industries, some forms of entertainment enjoyed by Canadians, some houses and farms of ordinary people, and she hoped to visit Chicago, where a British trade fair was being held in July, but it should be by the royal yacht, *Britannia*. Queen Elizabeth would like to entertain on the yacht – fifty-six could be seated for dinner and up to two hundred inside or up to three hundred on open decks for receptions. We must remember to allow time for hair-dressing, changing frocks, resting, with one day free each week, if possible, and Sunday, except for church. Representatives from industry, labour, and government should be presented. Massey suggested that his country place, "Batterwood," north of Port Hope, would be a good place for a day or two of rest. Finally, Prince Philip hoped that by early December I would be able to come to the Palace with an outline of the plans.

Massey felt very strongly that the ceremony for the Seaway opening should be at the Canadian end, near Montreal, and that the important part taken by Canada in planning and building this great international project must not be overshadowed by United States officialdom. I kept this in mind. Massey made his comments after I told him that I was going to see Livingston Merchant, the U.S. ambassador, that afternoon and that I was going to Messina, New York, the following week to meet Americans coming from Washington to discuss plans for the opening. I promised to make no firm commitment of any kind until it was cleared by our government.

I realized it was of prime importance to keep everyone informed in this extensive and complex program not only for the Seaway opening but also for the entire tour. To cope with the problem, I suggested to the PM that a committee of cabinet ministers composed of representatives from each province be appointed, and that I would report to them periodically on the progress of planning. This was done and George Hees, Minister of Transport, was chairman. In addition to this top-level group, I saw the PM at 8:30 a.m. several mornings of each week when he was in town to get his approval or decision on certain points. And similarly, I called on the Governor General between 5:00 and 6:00 p.m. and gave him progress reports. This was important because I sensed that the relationship between him and the PM was not exactly cordial. I recall that when I told the PM that our plans included a short weekend stay by the Queen and Prince Philip with Massey at his country house, the PM queried this and wondered why it was necessary. Early in the planning I learned that a cabinet minister or the Prime Minister must be in attendance upon the Queen throughout the entire tour. As a result, a particular minister coming from his own area of the country was detailed for this duty and I prepared a roster showing date and place where each was to take over from his predecessor. I sent each a detailed letter of instruction and there were no slip-ups.

As soon as my appointment as commissioner in charge of the royal tour was announced, I wrote identical letters to each premier and lieutenant-governor to introduce myself and say that in the near future I would be visiting them. However, in the meantime would they select and advise me of the name of a provincial co-ordinator with whom I could discuss plans of the visit to their province. The co-ordinator would keep the premier informed and also acquaint me of his wishes. This arrangement worked extremely well, and the co-ordinators appointed were cabinet ministers or senior civil servants. It was of some advantage that only a few weeks earlier I had met most of the premiers during my civil defence discussions. I asked H.F. Fever to act as my deputy and kept him in the picture, so that if I became a casualty he would be able to take over.

Early in November, I concluded that I needed expert advice from the

Navy and Air Force with reference to air travel and movement of the royal yacht, in which the royal party would travel from the lower St. Lawrence to the head of the Great Lakes. The Chief of the Naval Staff came to my aid and loaned me Lieutenant-Commodore S.M. King, and the Chief of the Air Staff did likewise and loaned me Flight-Lieutenant D.T. Thompson. These two officers were meticulous in their planning and advice. In addition to the naval and air help, I had the services of Frank Collins of the Canadian National Railway to advise me on railway equipment, allotment of accommodation in the train, time schedules, and suitable places for the royal train to park for the night.

During November, I had authority from the committee of cabinet to engage Major-General M.L. Brennan, recently retired as Adjutant General, as my chief executive officer. I was also able to hire Major J.M. Barry, a recently retired Army Service Corps officer, to arrange the advance booking of accommodation. I also borrowed Major E.G. Hession from the Army to organize ground transportation. There were no suitable motor cars available so I contacted the heads of Ford, General Motors, and Chrysler and they were most co-operative by each modifying one sedan, by raising the rear seats, and by removing the usual closed top and providing a plexiglass top that could be quickly attached in the event of rain. These cars were "leap-frogged" across the country in an RCAF transport plane – the C119, called a "flying boxcar." Cars for the staff to be used in each community were provided gratis by the dealers for the three companies. It was a very generous gesture on the part of the companies. When we arrived at a town or city, whether by boat, air, or train, the cars were lined up and ready, with the drivers carefully briefed, to drive through the streets, invariably lined with cheering, flag-waving crowds.

Security and crowd control were of immense importance. Here again I had excellent co-operation from the Commissioner of the RCMP, L.H. Nicholson, who loaned me Assistant Commander D.O. Forrest. His task required not only expert knowledge but also tact and patience, and he had these qualities to an exceptional degree. He had to deal with provincial and municipal police forces, and sometimes they were jealous of their own responsibility and resented interference from the RCMP. Douglas Forrest understood and sympathized with their attitude, and the result was a friendly and co-operative security system. Dealing with the U.S. Secret Service was a problem in itself. The program for the armed forces included provision of guards of honour, presentation of colours, and memorial parades, involving all three services, Navy, Army, Air Force, and each wanted its fair share of these events. To take care of these activities, I had the loan of Group Captain Gordon Richards of the RCAF, and with his help we managed to keep a proper balance.

The Queen, ever mindful of the fact that she would be here as our Sovereign, wanted as many Canadians as was practical to be members of her household. She therefore asked that we nominate a member from each of the armed forces to act as her equerries on this tour. This was done. Public relations was of prime importance. To find a suitable person to take on this task, I consulted members of the press gallery and some radio and television people. As a result I secured the services of R.C. MacInnes, Chief of Public Relations, Trans-Canada Airlines, who had previous experience with royal tours. Large numbers of American newspeople would be present, and throughout the tour there would be representatives from foreign countries. On this occasion, unlike former royal tours, we would also have to provide for television coverage, so facilities for transmission had to be arranged, which in a good many cases involved considerable expense.

The various means of transportation the Queen would use in moving across Canada on this tour created a problem for the movement of a large number of newspeople, some of whom would be travelling throughout the tour and others for only certain parts. To provide for the movement of these persons, aircraft were chartered by us but the cost was borne by those using the facility. In each province, a public relations man was appointed by the provincial co-ordinator, and my press staff worked in close co-operation with the provincial men. Over five thousand cards of accreditation were issued from our press section, and each recipient had to be cleared from the security standpoint.

The preparation of the program had to be an ongoing project. As decisions were made with reference to timings and events, these had to be passed to someone who would have them typed and in proper order. Thus, when the final item for the last day of the tour was settled, a book containing all information would be ready to go to the printer. The book totalled 347 pages, and it had to be printed in both French and English. Here again, I was fortunate in having the loan of Anne Corbett from External Affairs. Mrs. Corbett had prepared the program for the Queen when, as Princess Elizabeth, she had visited Canada. We had the task also of preparing suggested guest lists for the dinners, lunches, and receptions being given by Her Majesty on the yacht. We would then send them to the

Palace with suggested seating arrangements. The lists as approved by Her Majesty were returned to us with printed cards and envelopes, which we completed, addressed, and mailed, and then kept account of acceptance or regrets. Mrs. Corbett, with assistants, also looked after this chore.

Still another responsibility was securing copies of speeches intended to be made by premiers, lieutenant-governors, and others in welcoming Her Majesty. Copies were sent to the Palace so that she would know what to expect and her secretary could prepare an appropriate reply. I looked after this business myself and it was a trial. Politicos could not be made to understand why a one- or two-minute speech – and I gently suggested they be kept short – should be prepared months before it was to be delivered. The speech the Queen was to deliver from Ottawa on July 1 was drafted in the Prime Minister's Office and sent to Her Majesty for her approval or alteration as she saw fit, and the same applied with regard to her short speeches at the Seaway opening and the power development at Cornwall.

The baggage-handling must also be mentioned. There were four and a half tons of it, belonging to the Queen and Prince Philip, the members of the household, the staff, and the media. If only one mode of travel had been used, there would have been little difficulty, but with the frequent switching from ship to air to train to road, very careful planning and checking were necessary. To do the job, I had Squadron Leader H.A. Brisebois, loaned from the RCAF, and right well he did it. Our trunks, bags, and boxes were either waiting for us at each destination or arrived immediately after our arrival.

We were able to put together the itinerary for the approval of the Queen by early December, and on December 10 I arrived in London. During the next two days I had discussions with Sir Michael Adeane, the Queen's private secretary, and other members of the household. On Friday the 12th, Sir Michael, Sir Edward Ford, an assistant secretary, and Mr. Butler and I lunched with the Queen and the Duke of Edinburgh in Her Majesty's private dining room at the Palace. We then spent most of the afternoon going over the proposed itinerary and program outline. I had taken a large-scale map of Canada with me and had the proposed routes across the country clearly marked. This was spread out on top of the grand piano in Her Majesty's living room, as I gave the "travelogue" lecture. Upon return to Ottawa, December 13, we at once worked out new timings and new itinerary, which incorporated the few changes requested by Queen Elizabeth. This I submitted to the cabinet committee for their approval. The Prime Minister returned from his round-the-world trip just before Christmas, and in early January he also approved the itinerary. It was issued to the press after arranging with Mr. Butler at Buckingham Palace that it would be made known in London and Canada simultaneously.

The date of the opening of the Seaway was to be June 26 and the ceremony at the dam near Cornwall was set for the 27th. One sticky point had to do with the Secret Service men, the President's bodyguard. Their representative at the meetings was insistent that a whole platoon of about twenty should never let the President Eisenhower out of their sight. This would mean all of them being on the royal yacht. The RCMP argued that this was not necessary, they would have a small detachment on board, and everyone on the yacht was cleared for security in any event. After much haggling, a compromise was reached. Six of the Secret Service men came aboard and enjoyed the four-hour trip in the *Britannia*. However, in the motorcade from St. Hubert airport to the site of the opening ceremony, a car filled with guards simply had to follow directly behind the car carrying the Queen and the President. Normally this was the position for the Duke of Edinburgh and Mrs. Eisenhower. To meet this situation, we had the Duke of Edinburgh and Mrs. Eisenhower's car precede Her Majesty and the President.

We were, in early April, in a position to do a dry run over the itinerary for the entire tour. Mr. Butler came from the U.K., and on April 8 we set out in a RCAF plane from Ottawa and did the eastern part of Canada in ten days. On returning to Ottawa from eastern Canada, we spent three or four days dictating, checking, sending reports to London, and clearing matters with the Prime Minister. Then, we did western Canada and the Territories in another ten days. The dry run completed, we spent seven weeks on the immense amount of detail that had to be worked out and checked.

At the suggestion of Her Majesty, I was appointed, for the time she was in Canada, her Canadian secretary. Thus I became a member of her household and thereafter was referred to and treated as such.

At 3:30 p.m., June 18, the Queen and Prince Philip, with their entourage of thirty, arrived at Torbay airport in St. John's, Nfld., and the historic six-week tour of Canada began. The first week was spent in Newfoundland and eastern Quebec, including visits to St. John's and Gander, Corner Brook, Stephenville, Schefferville, and Sept-Iles on the north shore of the Gulf of St. Lawrence, where the *Britannia* awaited the royal party late Sunday afternoon. Then followed visits to Gaspé, a voyage up the Saguenay to Port Alfred, visits to the towns in that area, and then back to the St. Lawrence and a busy day at Quebec City and another at Montreal on June 25.

The second week started with the Seaway ceremonies on June 26, when the Queen and her husband and Prime Minister and Mrs. Diefenbaker met U.S. President and Mrs. Eisenhower, who had flown up from Washington that morning. The event was held at St. Lambert, across the river from Montreal, and was over by noon. The official party boarded *Britannia* and had lunch on board, as the yacht sailed eastward through the first three locks. All the guests took leave at about 4:00 p.m. at the Lower Beauharnois lock, and helicopters took them back to St. Hubert, from whence they departed for Washington and Ottawa. We had a quiet dinner and evening as the yacht anchored in Lake St. Francis.

The next day, June 27, Vice-President Nixon met the Queen and a short ceremony was held at the centre of the power dam near Cornwall. The 28th was a delightful, restful day cruising through the Thousand Islands and up Lake Ontario to Toronto, where the 29th and 30th were two very busy days, ending with a flight to Ottawa from where Queen Elizabeth broadcast an address to the nation on July 1. The following two days were spent in central Ontario and the Muskoka area. Early in the evening of July 3, the party again boarded *Britannia* and a pleasant and restful July 4 was spent sailing down Lake Michigan to Chicago.

Monday, July 5, was a busy day seeing the British trade fair and the sights of the Windy City, but Tuesday was a day for relaxation as we sailed up Lake Huron to Sault Ste. Marie. Wednesday was spent at the Soo, with a visit to the Algoma Steel Mills, and during the night we cruised up Lake Superior to Port Arthur. Thursday was divided between Port Arthur and Fort William with a departure for Calgary. Thursday ended the third week of the tour.

The fourth week, from July 10 to July 18, was spent in British Columbia and included a three-day holiday at a fishing camp at Pennask Lake. Travelling through the Rocky Mountains was done mostly by train to permit the royal party to enjoy the majestic scenery. On July 18 we arrived at Whitehorse, and for the following week made short calls at many small, mostly mining communities in the North, spent a day in Edmonton, most of a day in Regina, and a very full day in Winnipeg.

From Edmonton to Winnipeg we had travelled by train and stopped at a total of fifteen smaller towns where great numbers had gathered from the neighbouring countryside. The Queen and Prince Philip would walk among them on the station platforms, chatting in a most friendly way.

From Winnipeg on July 25 we flew to Sudbury for a visit to the nickel mines and thence to the RCAF station at Trenton. Here, after a word of welcome from the mayor, Ross Burtt (an old friend of mine), Her Majesty and the Duke of Edinburgh, with personal attendants, had a two-hour drive to Vincent Massey's home "Batterwood," north of Port Hope. Here ended the fifth week. The royal couple enjoyed a quiet weekend at the country home until Tuesday, July 28, when the party flew to Fredericton, N.B. From that day to Friday, July 31, the party visited the larger towns in Prince Edward Island and Nova Scotia, to arrive at Halifax at 6:00 p.m., Saturday. August 1 was the final day, and it had been intended that the royal party would board *Britannia* after a state dinner and depart for northern Scotland. However, this plan was altered and it was decided that they would return by air directly to London after the state dinner, emplaning at Shearwater Naval Base. This was arranged, and a few days after Her Majesty arrived back in London, the reason for the change in plan was made clear. A statement issued from the Palace announced that Elizabeth was *enceinte* and would not have further public engagements. In due course a male child was born, who was named Andrew.

I received many letters of commendation on the success of the royal tour and I passed copies on to the appropriate members of my staff, who really deserved the credit. One that particularly gratified me, which I passed on to Rod MacInnes, was from the Canadian Press.

Dear General Graham:

The Board of Directors of The Canadian Press met in Winnipeg last week, and I thought you would be interested to hear that particular reference was made at this meeting to the excellent arrangements you made for the press across the country in connection with the Royal Visit.

Those of us who have been in the newspaper business for a long time realize the unending difficulties faced in the organization of a tour like this, and it is not often we are able to write with such sincerity and unanimity after the event has taken place.

I hope you will pass along our thanks and our congratulations to those of your officials who particularly assisted the press in various parts of the country.

With kind personal regards,

Sincerely,

Charles H. Peters

Of all the questions asked with reference to the tour, certainly the most frequent are "What is the Queen like? Has she a sense of humour? Does she get easily upset?" From my experience I can say without reservation that Her Majesty indeed has a very fine sense of humour. When she is with her family and the members of her household and staff and at informal gatherings, she can be as merry as the best of us. I am sure that she must have been upset and annoyed and bored and very tired on many occasions during this long six-week tour, but never did I see any evidence of it. She is dedicated to the task of carrying out the duties that go with a constitutional monarchy, and she does so without complaint or reservation, ever ready to go beyond the call of duty. But despite the weight of official affairs, the Queen retains in the highest degree the finest attributes of a wife and mother.

Two incidents occur to me. In Winnipeg we drove through the business district on our way to the railway station. Here a picture was to be taken of the members of the household with the Queen and Prince, and there was quite a long delay in her appearance. It seems that in passing a large department store and from her car, she was able to see over the heads of the crowd a display of children's dresses in the store window. She liked the looks of one and sent her dresser to the store to buy it for little Princess Anne! On another occasion, my wife and I were in London, and when the Queen heard of it, she very kindly asked us through her secretary to come to the Palace for a short call. This we did and enjoyed reminiscing about the lengthy visit the year before. My wife inquired about the baby, Prince Andrew, and laughingly said we thought he should talk some day with a bit of a Canadian accent! The Queen laughed and said he was fine and growing fast. We left the drawing room where we had been chatting, and as we were going down a corridor with a secretary on our way out, Her Majesty came out of a side door with the baby in her arms, a proud mother, to show us wee Andrew. "Isn't he a lump?" she said with a chuckle.

14

THE TORONTO STOCK EXCHANGE

One night in early 1959, as I was sitting in the Ottawa air terminal waiting for a flight to the West connected with my planning for the royal tour, I met an old friend, Beverley Matthews, QC, a senior partner of the long-established legal firm of McCarthy & McCarthy in Toronto. After the exchange of the usual pleasantries, he asked me when I would be finished with my present task. I said I hoped in about six months' time and I was then moving to Oakville. He said he had something in mind if I would get in touch with him as soon as I was free.

In October, 1959, I had wound down all the business connected with the royal tour and had bought a house and moved to Oakville. I telephoned Bev, made an appointment, and learned what he had in mind. It was to accept the position of president of the Great Lakes Waterways Development Association, a consortium of a number of the larger users of the St. Lawrence Seaway, including the Steel Co. of Canada, Dominion Foundries and Steel Co., Canada Steamship Lines, Ontario Paper Co., and others. The presidents of these companies, William Scully, Frank Sherman, Rodgie MacLegan, and Arthur Schmon, were the directors of the association, and the purpose of the group was to watch over and really to "lobby" for what the users of the new Seaway considered to be fair rates or tolls on the movement of their vessels and cargoes through the new waterway system, which extended from Montreal to Lake Erie. Instead of each company acting alone, they had joined hands to present a common front to the Canadian government and, as president of this consortium, I would act for them. It was as though I had a retainer from each of them, my annual salary representing my fees. There was a similar organization of American users of the Seaway to deal with the U.S. government.

A dictionary defines "lobbying" as "an attempt to influence legislators in favour of some special interest." This is correct, but in doing his work, the lobbyist must assemble and present facts he hopes will convince the legislators that a certain course of action is in the interests of the country as a whole, and not only of special interest to his client. This task required considerable research, detailed study, and comparison of the costs of moving various types of freight by water, rail, and road, and, of great importance, the influence of existing government subsidies, both apparent and hidden, in support of these methods of transportation. I spent a busy few months on research and met several times with our American counterparts in Montreal, Cleveland, and Washington. It did not take me long to discover that either large government subsidies or hefty increases in existing tolls would be necessary to meet interest and maintenance charges, let alone make payments to retire the capital cost of the Seaway. There had been a massive miscalculation in estimating the cost of the project. It was estimated to cost $300 million, whereas it eventually cost over $600 million. Equally miscalculated had been the expected revenue from tolls as first established. Furthermore, operating and maintenance costs were almost four times the amount originally estimated. The original plan, as presented by the St. Laurent government to the Commons, indicated that revenue would be sufficient to pay interest, carrying charges, and operating and maintenance costs, and to retire the cost of building the Seaway in fifty years!

The Diefenbaker opposition and the government were seriously at fault in not delving deeper and more carefully into the financial implications of this great project. While my association and its American counterpart were urging their governments to refrain from raising the tolls, there was an equally strong lobby by the tidewater ports on the east coast and the railways and related labour groups to induce the governments to fix the rates. They wanted to set them at a level that would meet the full cost of the operation of the Seaway and the retirement within fifty years of the capital cost of building it, as had been originally planned.

On behalf of my clients, after months of research, I prepared a brief in support of our contention that it was in the public interest to keep the tolls at their present level. Then I presented it to the Minister of Transport, George Hees. This was late in 1960, and whether it had any effect on government policy I do not know. However, during my time as president of the association, the tolls were not raised and, in fact, were removed on the Welland Canal portion of the waterway. This was a political move on the part of the Diefenbaker government just before the election, and the tolls were reimposed by the Pearson government shortly after their election. Having done the necessary research, I found that the ensuing work with the association would not keep me as busy as I would like to be. I was very happy when Bev Matthews and Senator Salter Hayden, QC, the two senior members of McCarthy & McCarthy, suggested that I move into their offices. While continuing to act for the association, I would be shown on the McCarthy stationery as an associate counsel. My name would appear along with John J. Robinette, QC, a leading counsel, and Dr. J.W. Fox, QC, an international authority on patent, copyright, and trademark law, "below the line" drawn under the list of regular members of the firm. I knew several of those members, and Salter Hayden had been a classmate of mine in my first year at Osgoode Hall.

Now, after twenty years away from my chosen profession, I was back on line and decided to specialize on the laws relating to tariff and taxation, a field that year by year was becoming more complex and abstruse, as our lawmakers tried, by legislation and regulation, to find ways to meet the politically popular but very expensive and ever-increasing cost of social welfare and social service programs.

But scarcely had I opened a book on that branch of the law in October, 1960, when my plans went awry. One day I was invited to lunch with Colonel J.G.K. Strathy. I knew him well because he had been my Director of Military Training, a tower of strength to me when I returned from Italy in 1944 and became Deputy Chief-of-Staff. Before the war he had been with Dominion Securities, one of the largest brokerage and underwriting firms in the country, and he was now the president of that firm. He had recently served a term as chairman of the Board of Governors of the Toronto Stock Exchange, the TSE. The purpose of the lunch was not only to have a friendly chat but also, to my great surprise, to ask if I would accept the position of president of the Toronto Stock Exchange.

It was common knowledge that the office had been vacant since the retirement of Arthur Trebilcock a year or more earlier, and that the provincial government, through the Attorney General, A. Kelso Roberts, had been suggesting, if not urging, the governors of the Exchange to find a person for the job who was not a member of the brokerage fraternity. At that time the public's image of the TSE left much to be desired. The provincial government was loath to get involved intimately in the operation of the Exchange. They wanted it to be a self-regulated, self-policed operation and the target of "the slings and arrows" of outraged investors. They believed that if an outsider were appointed as president, and was not simply a figurehead, some existing practices and policies might be changed that would make it apparent that the TSE was being operated in the public interest and not solely to the benefit of its members, as many people thought was the case. The Ontario Securities Commission had been formed a few years earlier, with George Drew as the first chairman. O.E. Lennox was in 1960 the chairman, but the Commission exercised few powers over the operations of the Exchange. Lennox had a very small staff, and was involved principally in the approval of prospectuses and financing proposals and arrangements by Ontario corporations.

I listened to Jim Strathy's proposal. My immediate reply was that I knew nothing about stock trading, that I doubted my suitability for the job, and that in any case I was already employed and looking forward to a few more years in my chosen profession. However, Jim was insistent that I meet the chairman of the Board of Governors of the Exchange. He explained that the executive vice-president of the Exchange, W.L. (Bill) Sommerville, was well versed in the technical aspects of the Exchange operation, and I would not find it difficult to learn quickly all that was required. A proven administrator was needed. He reminded me that I once had under my command tank regiments, artillery units, engineers, signallers, and others, but had not been a qualified operator of any of them, although I knew what each should do! I agreed to meet Eric Scott, the chairman of the Board, and George Gardiner, the vice-chairman, at the Exchange the following day. On returning to my office after the rather lengthy lunch, I spoke to Bev Matthews and asked him what he thought of the proposal. He said Strathy had spoken to him in confidence and asked if he thought I would be a suitable president of the TSE, and Bev had been good enough to say that I would indeed be a good choice.

The next afternoon I met with Eric Scott and George Gardiner in the vacant office of the TSE president. I had never previously met either of them and later learned that when Jim Strathy reported to the Board that they should seek the services of General Howard Graham, Eric had said, "And who the hell is General Howard Graham?" We had a useful discussion as to what my duties would be. I made it clear that I had no desire to be a figurehead. They assured me that I would have the co-operation of the Board in any changes in personnel or policy that were felt to be necessary to improve the functions and image of the Exchange, and they proposed a salary I considered to be generous. I told them that I was still under contract to the Great Lakes Waterways Development Association and would like a few days to think about their proposal. They agreed. The Waterways chairman was good enough to say that he and his associates regretted my leaving but felt that I could make a more important contribution to the public by going to the Exchange. I phoned Leslie Frost, the Premier, for his reaction to my taking on what I appreciated would be a challenging and onerous position. He was emphatic in his opinion that I should take it. So it was that I vacated my office at McCarthy & McCarthy and, on January 1, 1961, moved into the office of the president of the Toronto Stock Exchange.

It took me but a few weeks to discover that, indeed, there was a vacuum to be filled. There was a lack of leadership, a laissez-faire attitude. William Sommerville, executive vice-president and a graduate of the law school, who had served in the RCAF, did an excellent job in supervising existing practices and operations. However, Bill did not conceive it to be his responsibility to suggest change while he served under a Board of ten governors representative of all types and sizes of brokerage firms. If the governors themselves did not see fit to bring to their meetings proposals for change, then nothing was done. There was a danger in this situation that the members of the Exchange would consider it to be a vehicle for their own advantage and lose sight of the fact that it was essentially an organization whose prime purpose should be to serve the public. Within a few weeks of my appointment, a conflict-of-interest case became a *cause célèbre*: Posluns vs. the Toronto Stock Exchange (1964).

Though the circumstances giving rise to the cause of action took place prior to 1961, it was not until May 14, 1963, that the case came to trial in the Supreme Court of Ontario before Mr. Justice Gale, later Chief Justice Gale. After a very lengthy trial, judgement was given on April 28, 1964, dismissing the claim of Wilfred Posluns for more than a million dollars in damages against the Toronto Stock Exchange. The judgement covered 137 pages of the Ontario Law Reports. Several of the most eminent counsel in Toronto were engaged, W.B. Williston, QC, and R.B. Tuer for the plaintiff; Arthur S. Pattillo, QC, J.F. Howard, and P.F. Jones for the TSE; and John D. Arnup, QC, now a Justice of Appeal, and George Finlayson, QC, for George Gardiner, chairman of the TSE's Board of Governors. It will be appreciated that, in addition to the customary duties as president of the Exchange, I was heavily involved with the Exchange solicitor, Allan Graydon, QC, and Pattillo and Arnup in the preparation of the defence and in attending the trial, which extended over several months, both as a witness and an observer.

Wilfred Posluns was a graduate of the University of Toronto and a young man of good character and unquestioned business ability. He became a member of the brokerage firm of R.A. Daly, a member of the TSE. To do so under a TSE by-law, he had to have the approval of the Board of the TSE, and this was duly given early in 1960. A year or two before that, Posluns and his two brothers had become associated with Dr. Morton Shulman in a partnership called Lido Investments. Dr. Shulman, as a sideline to his medical practice, had been personally interested for some years in investments, and was one of the first persons in Canada, I believe, to become involved in the trading of what was then called "puts" and "calls" and which were later more accurately called "stock options." The option most frequently traded is a "call," which means that the buyer of the option may call on the seller of the option to deliver a given number of shares of a certain stock at a certain price at any time within a stated period. This means that the seller of the option must either own the stock or be prepared to buy it at the market price if and when the option is exercised. Hence a dealer in options must have available a large sum of ready capital at hand. It was probably the need for availability of this capital that induced Dr. Shulman to enlist the support of the three Posluns brothers in forming the partnership of Lido Investments, in which Shulman had a half-interest and the Poslunses among them the other half.

A man named Lynch in Peterborough became interested in the stock option business. While Posluns was with the Daly brokerage firm in 1960, Lynch sold more than three hundred stock options with the Daly firm as his broker. Daly sold these same options to Lido, in which Posluns and his brothers had a half-interest. Lido in turn sold some or all of these options to an immediate purchaser who was willing to pay a higher price. This latter fact was not disclosed to Lynch. When the history of these deals became known to the governors of the Exchange shortly after my arrival, they took a serious view of Posluns's position in benefiting financially as a member of the Daly firm in the fees charged to Lynch and at the same time making a profit as a partner in Lido. Under the by-laws of the Exchange, the governors had approved the association of Posluns as a member of the Daly firm and had the authority to withdraw that approval if, in their opinion, the person had acted improperly. After a hearing with Posluns and his solicitor and with our own solicitor, Allan Graydon, present, the Board did withdraw such approval and in effect denied Posluns the right to be an associate of any member firm of the Exchange. In my view the severity of the penalty was a notice to all others that full disclosure must be made of all relevant facts in dealing with customers. As for Posluns himself, Mr. Justice Gale said in his judgement that the plaintiff "is an extremely able man who bears a fine reputation" and that one must assume that he did not appreciate the seriousness of the type of action that he became involved in. This action was of great importance to the Exchange and all of its members. At various periods from February, 1961, to April, 1964, when judgement was given, it occupied much of my time.

Shortly after taking office, I went to New York to acquaint myself with the operations of the New York Stock Exchange and the American Stock Exchange. I learned much that would be of assistance to me in proposing to the governors such innovations and changes in procedures as the introduction of a stock-watching unit in our staff organization. Its duty would be to note and report and determine the reason behind any unusual and unexplained movement in price or volume of a particular stock. I also became familiar with and a strong supporter of the term "democratic capitalism." During my years as Exchange president, I preached the gospel of democratic capitalism at luncheon and dinner meetings of a range of associations – Canadian Rotary, Kiwanis, and other service clubs, small investment clubs, and any other group that would listen to me. I would describe the organization and function of the Stock Exchange, and then explain the different types of securities, e.g., debentures, bonds, preferred and common shares, warrants and options – emphasizing that the higher the interest rate, the greater the risk, and that the "odds against" were high in buying speculative stocks like unproven mining issues unless one was able and prepared to lose his stake. I used some strong words to bring home my message. I described unscrupulous mine promoters as "cheap tin-horn tycoons" and "jack-als." I did not hesitate to be critical of some of the members of the Exchange when I said that "many brokers and their employees seem to forget that the Stock Exchange is an institution which should be operated in the interest of and for the benefit of the public."

But I emphasized again and again the responsibility of an investor in putting his savings in an unproven stock. My mission was to educate the public. I considered this to be a function and a duty of stock exchanges. By educating and lecturing, I was trying to induce more of the public to invest in Canadian companies. I encouraged the formation of small investment clubs and spent many an evening explaining to them the desirability of Canadians owning their own natural assets and having shares in the companies that would process our natural resources. That gospel remains as relevant today as it was a quarter-century ago.

In 1961, the directors of the Toronto Stock Exchange learned that Eric Kierans, the president of the Montreal Stock Exchange, while abroad and perhaps with some justification, had referred to the Toronto Exchange as a gambling casino where most of the business was in the trading of highly speculative penny stocks, whereas the Montreal Exchange dealt in blue chips, proven, dividend-paying shares in established companies. My Board thought it would be a good idea if I went abroad to shed more and better light on our activities. At the same time, unemployment in Canada was at a high level. In addition, the Diefenbaker government and J.E. Coyne, the Governor of the Bank of Canada, were having an open disagreement on the bank's policies. Both of these factors were having an adverse effect on foreign investment in Canada. To cover these two subjects, the types of listings, and the volume of trading done in Toronto compared with other exchanges in Canada, I wrote a brochure of some sixty pages and had it printed in English, French, and German and took copies with me on my journey. George Hees was then Minister of Trade and Commerce, and with his support I wrote to the ambassadors and trade commissioners in the cities I planned to visit. They gave me most willing help in arranging appointments with the various stock exchanges and senior bank and government officials.

I flew first to Ireland and visited the Dublin Exchange. I found it very small, open two or three hours a day, and dealing in the shares of mostly local companies. Then I flew to Glasgow and attended a trading session there and said a few words to the members, as I did at all of my calls. I then spent a day in Edinburgh. At the Exchange in Liverpool, I said a few words and distributed my brochures. One member wanted to know when the unfavourable balance of trade that Britain had with Canada would be corrected. I replied that while I could understand his concern, the reason for the imbalance was that most of the British purchases from Canadian companies consisted of raw materials, while Canada's purchases from Britain were entirely of processed goods, which gave employment to many times the number of people who were engaged on our side in harvesting wheat. Hence, while the dollar value of our trading was in our favour, the employment factor was many times over in their favour, and they were getting the best of the deal. There were no further questions!

I then visited Manchester, Birmingham, Portsmouth, and London. All except London were small area exchanges. In the world of stock exchanges, the New York Stock Exchange ranked first in volume and value of stocks traded, with London, Tokyo, the Paris Bourse, and Toronto following in that order. Lord Ritchie of Dundee, the president of the London Exchange, had me as the guest at a governors' luncheon, and I spent several hours seeing their active trading and discussing their practices. Their members, in common with members of all exchanges in Britain and on the continent, were much interested and sometimes critical of our methods of financing the exploration and development of our mineral resources. One member complained: "I understand that only one-third of the money paid by investors in treasury shares is actually used for company purposes, and the balance goes to underwriters, promotion, and sales personnel." I had to admit that this was true in some cases but steps were being taken to correct the situation. After the London Exchange closed, I went around the corner to call on the head of the Bank of England. I was admitted by a doorman dressed in the centuries-old style of pink frock coat and silk hat!

From London I went across the Channel to The Hague, Brussels, Dusseldorf, Frankfurt, Munich, Zurich, and Paris. Our ambassadors and trade commissioners had set up appointments for me, and in many cases the latter went with me on my visits to the exchanges and banks. The president of the National Bank of Germany was particularly interested in the Coyne-Diefenbaker row. Many times I was asked about the future of cities and areas where foreign funds had been invested in office buildings and housing developments. As I had visited all parts of the country, I was able to identify the projects they were referring to and to make what I hoped were intelligent comments about their properties.

At one bank in Frankfurt, the president was critical of our large unemployment figures. I admitted that they were unusually high, but I added that one reason was the increasingly high numbers of women who were entering the work force which, I understood, was not the case in Germany. I said, "When I came into your banking hall this morning, I saw only men meeting and waiting upon the customers. Only an occasional female working at a typewriter was to be seen. When I entered the elevator to come up to your office, a man was operating it. When I left the elevator, a man was at the reception desk. In Canada now, one seldom sees a man in the banking hall or operating an elevator or at a reception desk. All of these jobs are taken by women, a great many of whom are married." My host was quite surprised at this invasion of the work force by females and was very critical of the practice!

One practice I found common in Europe was the use of a central depository for shares. Under this system a purchaser, instead of receiving a certificate for his shares, had an entry in a book somewhat as a depositor receives an entry in a bank book when he deposits money in a bank. The record of the share purchase is made in the books of the central depository, and hence the cost of issuing new certificates every time shares were traded was obviated. The bother and expense of having them in one's possession or in a safety-deposit box was saved. I was told that in 1940, when the Germans occupied France, they imposed the central depository system on the Paris Bourse, and it was found to be so satisfactory, as it had been in Germany for many years, that it was retained in Paris after the Germans left. I saw the advantages of the system and promoted the idea on my return to Canada. I found much opposition, particularly from the trust companies, which were loath to lose the fees they received for the issue of new certificates every time shares, whether one or one thousand, changed hands. Many brokers felt that their customers would not be satisfied with having simple entries in little books instead of owning beautifully engraved share certificates with their name upon them.

I recall speaking to a luncheon meeting of the Trust Companies Association on the merits of a central depository and predicting that within five years the securities industry in Canada would surely make a move in this direction. I was not far wrong. Such a system was inaugurated in 1970 by the incorporation of the Canadian Depository for Securities Ltd., a company owned by banks, trust companies, brokers, and investment dealers. While some securities were ineligible for deposit and some investors failed to use its facilities, it nevertheless had a quite phenomenal growth. The company in its 1983 report showed 1,162 securities eligible for deposit with more than $13 billion in value of securities on deposit. Another innovation I recommended to the governors of the Exchange was the appointment of two non-brokers to the Board. I felt that such action would further improve the image of the Exchange and indicate more clearly to the public that the TSE was not a closed corporation operating only for the benefit of the brokers. I was happy to see my recommendation acted on shortly after I retired.

In early November I returned to Toronto, tired after much travelling but certain I had delivered the message that the TSE was a responsible institution and that there were good reasons to invest in the future of Canada. In early 1963, I made somewhat the same sort of peregrination to the Far East, visiting New Zealand, Australia, Hong Kong, and Japan. On this trip, happily, I was accompanied by my wife. In New Zealand and Australia one receives a particularly warm welcome because the folk there appreciate that you have made an out-of-the-way journey to see them. When I arrived by air at Sydney, Australia, I was greeted by the president of the Stock Exchange and a dozen or more press and television representatives with camera crews. A room had been arranged and we had quite a lengthy interview, which was used on television that evening and in the daily press the next morning. I was told that the state treasurer would like to see me, and the Exchange president had a full schedule of appointments, lunches, and dinners.

The subject that interested the treasurer and others the most was our method of raising capital for the exploration and development of our natural resources. They particularly wanted to finance such activities in Australia as much as possible with local funds and not sell out to foreign investors. I was able to tell them of our methods, being candid about the weaknesses. I emphasized that being a close neighbour to the United States and in too much of a hurry to find and develop our great store of irreplaceable natural assets, we had sold out to our neighbours to the south vast quantities of those resources. I do not know whether our discussions proved useful to my hosts, but certainly they were of interest to me and led to a closer relationship between our respective financial communities than in the past.

I spent some time in Canberra, the capital of Australia, with Evan Gill, our High Commissioner and an old friend of mine. He had arranged for discussions with various officials of the federal government.

From there I flew down to Melbourne and went through the same exercise as in Sydney. From Melbourne we flew up to Hong Kong, a very important financial centre. Much building of apartment houses and business blocks was going on, and much of it still in a most primitive way. I recall seeing women carrying baskets of scree on their heads from excavations for the erection of buildings in the New Territories. At one of the largest banks I asked the president who was financing these projects. He replied that his and other banks and trust companies were supplying the capital. When I questioned the validity of the security, when the British lease on the area would expire before the end of the century, he said the apartment houses would pay for themselves long before the lease expired. He explained that in most cases the buildings were occupied by three families, each having eight hours per day tenancy when they slept, ate, and did their washing. The other sixteen hours were spent at work! The large East Indian population in Hong Kong was principally engaged in retail business. The Chinese and British, who were employed in banking, brokerage, and business occupations, seemed to be perfectly integrated and on equal footing. The foremost club in Hong Kong had members who were both Chinese and British. On the Stock Exchange and in the banks, Chinese held positions of equal importance to the British.

We flew from Hong Kong to Tokyo and were received with great warmth and hospitality by the officials of the Stock Exchange, where the trading floor was about a half-acre in size and the babel of trading was deafening. It was interesting to see how the Japanese were accomplishing what I had been trying to do in Canada – encourage average citizens to be interested in and invest in their own securities. I was given a wooden box, very attractively made, about eight inches high and four inches square, with a lockable hinged top with a slit in it. Japanese characters were painted around the side with the message "A Window to Happiness." Such boxes were essentially "piggy banks." They were issued by brokerage houses to families of modest means who would place them on dining-room tables or other handy spots. Members of the family were encouraged to put any spare cash in the locked box. Periodically a collector arrived, unlocked the box, gave the householder a receipt for the cash, and with collections from a great number of such people bought shares in companies of the broker's choice, unless the box-holder specified a stock preference. The practice encouraged great numbers of low-income workers to put their spare cash into the support and growth of Japanese companies, probably companies in which the investors were working.

It is a rare occasion when a foreigner is entertained in a private home in Japan, but the president of a brokerage firm, when I called upon him, asked if my wife and I would come to his home for lunch the following Sunday. I gladly accepted, and on a dreary snowy March morning we were driven to the outskirts of Tokyo to a typical Japanese bungalow, a large, frame building on what seemed to be a small estate. In the entrance hall were cloth slippers. We shed our shoes, donned the warm, clean footgear, and were led to the sitting room, furnished mostly with wicker items, except for a wood-burning stove. Our host introduced us to his lovely wife, who did not speak English, and half a dozen other guests, some of whom did speak our tongue, including his brother, an Oxford graduate. After a glass of light wine, we went to the dining room, but we changed slippers before entering. The room was beautifully panelled. At one end was a large open dresser of several shelves, and these were covered with beautifully dressed dolls. The table was Japanese style, low, so that we sat on cushions on the floor with legs crossed or extended under the table. The food – many courses of soup, fish, cereals, and sweets – was easy to manage with spoons and chopsticks, which we had learned to manipulate. The whole visit was a delightful occasion and one that our ambassador said was seldom experienced by a foreigner.

Little time was spent on the social circuit. Each day for a week was filled with appointments – visiting brokers, bankers, factories, and the Stock Exchange. I marvelled even at that time at the modern methods and equipment that my hosts were obviously proud to show me. Certainly their progress since the end of World War II had been phenomenal. They were a people dedicated to their nation and its future. In the offices, in the factories, in the shops, I saw only Japanese. Their efforts and national pride were not diluted by any influx of foreigners or minority groups, whose rights to their customs, languages, and religions had to be protected, usually at the expense in dollars and the productivity of the majority.

In late March I was back in Toronto, where Bill Sommerville and the TSE staff managed things during my absences. Unquestionably, my journeys to Europe in 1961 and to the Orient in 1963 did much to improve the image of the Toronto Stock Exchange, opening lines of communication that benefited the financial communities of the countries visited and of Canada. Thereafter we had visits from the presidents of both the Melbourne and the Sydney exchanges and officials of the Tokyo Exchange and representatives from brokerage firms who hitherto had been visitors to New York and London but had bypassed Canada.

During my absence in 1963, the preparation for the Posluns claim had gone ahead. Almost at once upon my return I was involved in aiding our solicitor and counsel in putting together the large amount of correspondence and records that would be needed at the months-long trial. Judgement dismissing the claim was delivered on April 29, 1964. The ink was scarcely dry on the Posluns judgement before the Stock Exchange became involved in another series of events that were destined to result in several important changes in our policy and operating procedures. I refer to the trading in the shares of Windfall Oils & Mines Ltd. In late 1963, the Texas Gulf Sulphur Co., after extensive aerial reconnaissance across the Canadian Shield, began drilling in a large area it had acquired in Kidd Township north of the town of Timmins, Ontario. Timmins was built and prospered to exploit the rich and extensive gold mines in the vicinity. But during the past half century, these deposits were being worked out and the town's future looked bleak indeed. Hence there was much interest in the activities of Tex Gulf, as the company was called. However, the company made no statement as to its success and played down any rumours that began to circulate in Timmins and among the mining and prospecting fraternity. In fact, a vice-president of Tex Gulf, as late as April 12, 1964, issued a statement that rumours of a strike were "premature and possibly misleading."

However, on April 16, the company acknowledged, through an article published in *The Northern Miner*, that there had been a discovery of sulphides of considerable magnitude. Two days later, Viola MacMillan purchased twelve staked claims adjacent to the Tex Gulf property and immediately sold them to Windfall Oils & Mines Ltd., a company controlled by her and her husband George. Once Tex Gulf announced the magnitude of its discovery, tremendous interest was aroused in other properties in that area. Rumours of further strikes of great potential resulted in a rush on the part of the public to make a fast buck or a substantial fortune. This had been the case for many people who bought for a few cents shares in great mineral producers like Lake Shore, Teck Hughes, and Kirkland Lake gold mines. Rumours began to spread that Windfall had a good chance of success. George and Viola naturally did nothing to discourage this speculation during the weeks following their acquisition of the properties near the Tex Gulf strike. On July 1, 1964,

Windfall started drilling, and within days unconfirmed stories began to circulate that the shares of the company were a good but still speculative buy. By July 7, the shares of Windfall were selling at over $2.00, up from fifty cents a week earlier. By July 13, they reached $5.00. The Exchange was very concerned about this rapid rise and about the fact that the MacMillans were evidently holding back any authoritative statement on what values the assays were showing from the cores that had been drilled. We were receiving no help from the Securities Commission in pressing the MacMillans for these assay results and requested the chairman of the Ontario Securities Commission (osc) to call a meeting of the Windfall owners (the MacMillans) and representatives from the Exchange and the Securities Commission.

As a result of this request, a meeting was held on July 14 in the offices of the osc. In all, ten people were present. We had threatened to suspend trading in Windfall until the company announced the result of the assays, and we hoped that the osc would support us in this action. Basically, suspension is a disservice to people who bought shares and would be "locked in." For this reason we were loath to suspend trading without the Commission's announced support and agreement. Perhaps we were wrong in taking this position, but the fact remains that the Commission did not encourage us and, in fact, rather discouraged us by issuing a statement to the effect that Windfall had not completed its assays of drill cores and that any stories of assay results were unfounded. The statement did nothing to dampen the wild trading and extravagant rumours. On July 21, the shares reached an all-time high of $5.60 a share. Not until July 30 was it announced that the assay of the Windfall cores disclosed little of value. The shares plunged to $1.00.

On August 13, Arthur Wishart, the Attorney General, with the support of the TSE, appointed Mr. Justice Arthur Kelly of the Ontario Court of Appeal as royal commissioner to report on "recent fluctuations" in the prices of the shares of Windfall and "circumstances and events" pertaining to sales, purchases, and other dealings in the shares. During the next twelve months the Kelly Commission made a complete inquiry into and study of the methods then in vogue with regard to the financing of new mining ventures and the role of the Securities Commission and Stock Exchange and its members in these activities. He pointed out the weaknesses in the system, as he saw them, and suggested changes that should be made. During the months of hearings by Mr. Justice Kelly, the Exchange staff was extremely busy conducting day-to-day business and attending the sessions to give evidence and to hear what others had to say.

During the weeks of Windfall's surge of trading, through meetings with members of the media and in speeches, I repeatedly emphasized that, by putting money into shares of a company that had not released the results of a core assay, investors were likely to "lose their shirts." I did not receive a single letter of criticism or complaint from investors about the Stock Exchange operations after the collapse of prices. No member of the staff of the Exchange was found to have traded shares in Windfall or been implicated in the rumours that caused the enormous volume of trading and the effect upon price of the shares. Almost twenty-nine million shares were traded in one day!

The Kelly Commission made a number of recommendations. From my point of view, one of the most important read as follows: "The administration of the Exchange and the conduct of its business, the supervision of the manner in which members conduct their business on the Exchange, and the initial investigations to determine if the conduct of members is in keeping with the principals governing the Exchange, are matters which should be left to the permanent staff, free from any interference by any members." This policy had been adopted and was operating successfully even before the Commission's report was issued.

As to Windfall Oils and Mines Ltd., it was delisted and as far as I know is now a dormant company. George and Viola MacMillan were charged with fraud in their dealings with Windfall shares. The trial was held before County Court Judge Harry Deyman and the MacMillans had as their counsel during the three-week trial Messrs. John J. Robinette, QC, Joseph Sedgewick, QC, and Kelso Roberts, QC, the former Attorney General. The defendants were acquitted and the Crown did not appeal. Viola, before that trial, which was not held until 1969, had been found guilty in another case involving "wash trading," i.e., manipulating the price of shares by artificially creating the appearance of trading activity; she was sentenced to nine months in Mercer Reformatory but served only a few weeks because of ill health.

I had been almost five years in the president's chair by the end of 1965. They had been busy years during which many changes had been introduced. Some were not welcomed by some of the members, particularly those who dealt in a large way in speculative stocks. As a consequence, a large part of the mining promotion and financing moved to the Vancouver Exchange. One day early in November, I attended a meeting of the Canadian Club and sat next to an old friend, Bishop Fred

Wilkinson. He had served as a private in World War I and had three times been awarded the Military Medal for gallantry in action. That morning the newspaper had carried an item that he had announced his intention to retire, and in talking to him I referred to this, saying that I was surprised and that he would be greatly missed. "Well, Howard," he said, "I have reached the allotted span of three score and ten, and although my health is still good, I feel that I should make way for a younger man. I can still do a few odd jobs."

Later that day I thought of his decision, and that evening discussed with my wife the possibility of my retirement in the near future. My arrangement with the governors of the Exchange was that I would continue as president for an indefinite period but might retire on giving six months' notice. Jean and I decided that it would be a wise and wonderful release if we could enjoy our home and garden on the shores of Lake Ontario in Oakville for a few years while our health was still good enough to enjoy it. So it was that next morning I wrote a personal note to Marshall Stearns, the chairman of the Board, asking them to accept my resignation effective June 30, 1966. This would allow more than the six months agreed upon to find a successor and would also enable us to complete changes in our by-laws that were still being considered as a result of the Kelly Commission's report. Jack Kimber, chairman of the Securities Commission, left that post and became the president of the Exchange. A better man could not have been found, so the TSE was in good hands.

The Board of Governors of the Exchange was exceedingly kind and generous. In addition to presenting me with a handsome retirement gift, they held a testimonial dinner at the Royal York Hotel on June 16, with the chairman, John Deacon, presiding, and members and employees of the Exchange in attendance. I was deeply touched and greatly honoured on this occasion to have as guests G.A. Gale, Chief Justice of Ontario, Arthur Kelly, Justice of the Appelate Court, Allan Lamport, the mayor of Toronto, my old friend Leslie Frost, then a vice-president of the Bank of Montreal and formerly Premier of Ontario, John Kimber, chairman of the Securities Commission, W.J. Gilling, Dean of St. James Cathedral and padre of the Hastings & Prince Edward Regiment while I was commanding in 1941 and 1942, and Lieutenant-General Guy Simonds, as well as past chairmen and members of the Board of Governors. The news media were well represented.

During my term of office, my relations with the media were, with few exceptions, most friendly and supportive in our efforts to effect reforms and improve the image of the Exchange – "our efforts" because, while I may have taken the lead in bringing forward proposals, the Board members took the necessary action, and they did it gladly and without pressure from me. The news reports on my retirement were generous. Fraser Robertson, the financial editor of *The Globe and Mail*, wrote on June 16: "Mr. Graham can look back with pride on the accomplishments of the Toronto Stock Exchange during his tenure of the presidential office. The Exchange during that period has responded positively to criticism and he leaves it an organization more able to adapt itself to developing public standards, more capable than ever of serving the multifarious requirements of its members and their clients, both large and small." Jack McArthur, the financial editor of *The Toronto Star*, wrote on February 18: "The first big push for reform came at the Toronto Stock Exchange. And the General was the boss. Now he's resigning – some say he'll be Governor-General. If he is, he won't be as important to Canada as when he was Howard D. Graham, President of the Toronto Stock Exchange, custodian of millions of dollars in Canadian investments. (That's meaning no insult to Governor-Generals). . . . Let's face it. One reason the Exchange moved was because it had an outsider of stature as its boss. Even the slightest suspicion that Graham might quit if he thought he was being pushed around by Bay St. must have been enough to send shivers through the financial community. The next stop might have been out-and-out government control."

I received many letters of commendation from old associates and many honours from groups. The Floor Traders Association made me an Honorary Life Member and gave me a splendid piece of luggage. To remind me of the great MacMillan fracas, they handed me a framed certificate for one share of Windfall Oils & Mines Limited!

15

NEW RESPONSIBILITIES

I was asked by Roger Teillet, the Minister of Veterans' Affairs, to accompany him as the Canadian Army Veterans' representative to ceremonies in France to commemorate the fiftieth anniversary on July 1, 1966, of the opening of the great Somme offensive of 1916. On June 24, my wife and I took the night train to Ottawa and were met on our arrival by one of my former staff sergeants in the Hasty P's, Gordon Way. He was then Director of Public Relations in the Department of Veterans' Affairs. Gordon had a car waiting, and we were driven to the RCAF waiting room at Uplands Airport where we joined Roger Teillet and his wife and other officials of the Department of Veterans' Affairs and of the Royal Canadian Legion. We left at once for Paris.

During the next two weeks, starting from Arras, where we made our headquarters, we attended short memorial services and laid wreaths at cemeteries in or near Vimy, Courcelette, Cambrai, Amiens, Albert, and Beaumont-Hamel, where the Royal Newfoundland Regiment had been almost completely decimated in the early days of the Somme campaign. In addition to the wreath-laying ceremony, flood lights, which would show to advantage the beautiful memorial statue, were dedicated. Field Marshal Earl Alexander of Tunis represented the British forces in these various ceremonies and usually I sat next to him. He said scarcely a word and never mentioned his term as Governor General of Canada. Perhaps he had forgotten it!

After the visits to the cemeteries in France, we flew from Paris to Rome and from there drove down to the large Allied cemetery near Cassino and from thence to Naples. The intervening day we drove across the Apennines to the all-Canadian cemetery at Ortona, situated south of the city on a bluff overlooking the Adriatic Sea. Here was the ground my brigade had fought over and held in late 1943. When I again saw the precipitous banks of the Moro River, the ravines and steep slopes covered with vineyards, many of the vines tied to interlaced overhead wire supports, I marvelled that we had ever been able to drive away the stubborn and experienced enemy forces. The cemetery was evidence of the immense cost in young Canadian lives. Unfortunately, our itinerary did not include the Canadian cemetery on the hillside outside Agira in Sicily, which I had visited with our ambassador, Pierre Dupuy, some years earlier. From Naples we flew to London and had a wreath-laying ceremony at the Cenotaph and at the Commonwealth Air Force memorial overlooking the site of the signing of Magna Carta at Runnymede.

While in London, Jean and I were honoured to be invited to the Queen's garden party and also to the royal enclosure, where the Queen and members of the royal family and household and a few invited guests had their tea and refreshments. As I lacked formal attire, I made an appointment with Moss Bros. to rent the suitable morning suit and grey topper. Jean had a suitable frock, shoes, and gloves, and was loaned a lovely garden-party hat by a friend we had known from our 1946-48 period in London. At Moss Bros., a century-old firm that rented out all sorts of uniforms, regalia, and clothing for the many ancient ceremonial occasions that are a part of English tradition, I joined many others on the same errand as I. The first thing one is asked to do when the attendant is told that you need the kit for the Queen's garden party is to go downstairs and fit yourself with a grey top hat. This done, you go back upstairs and in a few minutes are measured and fitted with the required striped trousers, tail coat, and grey waistcoat. Black shoes, white shirt, and grey tie are also required. A specially designed suitcase neatly takes the hat, suit, and accessories. I paid the equivalent of about twenty-five dollars rental and was bowed out with one last item added, a nicely furled umbrella! It is not a disgrace to be fitted out in this way, and when I was being addressed as "Sir," others were being styled "Your Grace" or "Your Lordship."

The next day Jean and I walked from our hotel across Piccadilly and the Green Park to the entrance to the grounds of Buckingham Palace and made our way to the roped-off area of the royal enclosure. Here, with a goodly number of others, we awaited the arrival of Her Majesty and His Royal Highness and other members of the royal family and of the household. They were making their way from the terrace at the west side of the Palace through the hundreds of guests gathered in the twenty-five acres of the Palace grounds. It was a beautiful July day, and though the umbrella was carried, it was not unfurled! One person with whom I had an interesting chat while waiting for Her Majesty was Lord Avon (Sir Anthony Eden), former Prime Minister. He was much interested in Canada and had visited the country on a number of occasions. He had served with distinction during World War I in the Rifle Brigade of the British Army and had been for a number of years Colonel-in-Chief of the Brockville Rifles, a regiment in eastern Ontario that had provided a number of reinforcement officers to my own regiment while I was commanding it.

Shortly after the Queen arrived, we were summoned by an aide, as was the custom, to have a few words with Her Majesty at the end of the marquee where her tea table was set up. How were we? Where had we been? Had I retired? Natural questions, easily answered and, needless to say, a great pleasure to Jean and me, who recalled our last informal visit with her at the Palace in 1961. A day or two after the garden party we returned to Canada and once again settled down in our pleasant home in Oakville by Lake Ontario for what we thought would be a long and quiet period of relaxation. But our quiet period lasted only a few weeks.

My period of retirement did not last long. Early in October, 1966, I received a phone call from one of the Prime Minister's secretaries. Could I come to the Park Plaza Hotel in Toronto that afternoon to see the Prime Minister? I said I would be glad to and the time was arranged. When I told my wife of the appointment she asked, as one might expect, "What does he want to see you about?"

I replied, again as one might expect, "I don't know but I'll bet it has something to do with next year's centennial celebration."

And so it was. I arrived at the Prime Minister's suite in the Park Plaza at four o'clock, and on being ushered into his living room found Prime Minister Lester B. Pearson relaxing on a sofa with his feet on the coffee table. He greeted me, friendly as always. "It is good of you to come in, General. How are you keeping? Enjoying a bit of a holiday after your five or six years at the Stock Exchange? I haven't forgotten your help during our troubles with Colonel Nasser." A few other pleasantries followed and then we came to the business at hand.

I learned that as next year, 1967, was the centennial of Confederation, the Queen and Prince Philip would be coming to Ottawa for a few days. Queen Elizabeth, the Queen Mother, would be visiting the four Atlantic Provinces, and Princess Alexandra, Mrs. Angus Ogilvy, and her husband would be visiting Toronto and the western provinces. In addition to these royal visitors, a large number of the heads of state with which Canada had diplomatic relations would be visiting Ottawa and Expo, the world exposition to be held in Montreal that year. Lionel Chevrier, then our High Commissioner in London, would return to Canada early in 1967 and be appointed Commissioner-General with responsibility for all visits. Lieutenant-General Robert Moncel, recently retired from the Canadian Army, was being asked to be co-ordinator of visits of the heads of state, and the Prime Minister was asking me to co-ordinate the three royal visits.

I told the Prime Minister that I appreciated the honour attached to such an assignment but would like to talk the offer over with my wife and think about it for a few days. I said it would involve commuting back and forth, probably each week, between Oakville and Ottawa, living in a hotel room most of the time, and once again leaving my wife alone for five or six days each week. Apart from this drawback, I immediately foresaw many problems in working out programs for three different people, each covering a different area of Canada. Again, as in 1959, there was the need to recruit a suitable staff for the task. The Prime Minister said he understood my attitude but that it would be a relief to him to have someone with my experience do the job, and, furthermore, "the people at the Palace" hoped that I would do it. I assumed that he was referring to Sir Michael Adeane, the Queen's private secretary, and other members of the household.

Jean was not keen about being again left alone for the better part of a year, but after a few days of thought we both agreed that, provided financial arrangements were satisfactory, I would take on the responsibility. I was worried about the effect of employment on my Army pension. At that time if a pensioner became a member of the civil service, the pension was abrogated while he was so employed. In my case, and I assume in many others, where one was engaged for a limited time and for a special duty, an order-in-council was passed that approved a contract between the government and me for one year at a fixed salary – the pay of a lieutenant-general plus out-of-pocket expenses. I went to Ottawa in late October. Office accommodation had been arranged for Chevrier, General Moncel, and myself in downtown Ottawa. Although Chevrier remained in London until early 1967, Moncel and I went about the business of gathering a staff and starting on our planning. I learned that as early as April, 1966, an ad hoc interdepartmental committee had been formed to consider policy and preliminary planning. Discussions had also taken place among the concerned parties. Of great importance to me was the appointment of Kenneth D. McIlwraith as my deputy. Ken was in the Department of External Affairs and for the past four months had been involved in the preliminary planning.

As I had done in 1959, I sent letters to the premiers and lieutenant-governors announcing my appointment and asking them to appoint provincial co-ordinators with whom I would deal. They were also told that the Queen would visit Ottawa and spend perhaps one day at Expo in Montreal; the Queen Mother would visit only the capital cities plus any city larger than the capital in the Atlantic Provinces; and Princess Alexandra and Mr. Ogilvy would visit only Toronto and the capital cities of western provinces plus Vancouver and Calgary. As one might expect, there were numerous requests from cities and towns other than those specified, and it was my task to write and say, "Sorry, there can be no exceptions to this policy." Not unexpectedly, there were two or three exceptions before the programs were finally settled.

Princess Alexandra and Angus Ogilvy were the first to arrive so we gave priority to developing their program, but at the same time we worked on those of the Queen and Queen Mother. Visiting all the provinces and territories, as I had done in 1959 to discuss suggested events, I was able to go to London on February 6, 1967, after securing the Prime Minister's approval, and meet the principals and their secretaries and sense their reactions to our proposals. The first day I had lunch at the Palace with Sir Michael Adeane and others of the household, and afterward we went up to the Queen's sitting room, where eight years earlier I had spread my map of Canada on the grand piano and with Michael had an hour or so with Her Majesty and His Royal Highness. For political reasons, I think, and also for the comfort and use of Her Majesty for entertaining, it had been decided before I came into the picture that the Queen and Prince Philip would visit Expo on Monday, July 3, arriving and leaving by the royal yacht *Britannia*. This was agreed to and so was the proposed program for visits to Commonwealth pavilions at Expo.

There were some difficulties in this matter. We wanted to avoid inconvenience to the public as far as possible and developed a route for these visits that would delay the admission of the public to about one-third of the Expo area by half an hour rather than an hour and a half. But this planned route through the Expo grounds provided for the first call to be made at the United Kingdom pavilion. If the Canada pavilion and the provincial buildings were first visited, it would mean a back-tracking that would be a waste of time and would delay the general admission. I explained all this to the Prime Minister and he agreed to the plan, but a few hours later I had a telephone call from the Prime Minister's principal secretary to the effect that Marc Lalonde, who was Pearson's French secretary, had told the Prime Minister that it would not be politically acceptable in Quebec if the Queen's first call was at the United Kingdom building and that I should change it. I am afraid that for once in the heat of battle I lost my patience and told the principal secretary that if the Prime Minister wants Lalonde to do the program to go ahead and I would be glad to be free of it. He cooled me down a bit and suggested I speak to Robert Winters, the federal cabinet minister responsible for Expo, and perhaps Bob would straighten things out.

I called Bob and told him the facts and he asked me to bring to his office in the Parliament Buildings the marked map of the Expo area with the route I had proposed. I did. After I gave him the details and the reasons I had so planned the route, he simply said, "Give me that map, Howard. I'll be back in a few minutes. Just wait." I did and in ten minutes he came back, having seen the Prime Minister, handed me the map, and said, "Okay, follow your plan." And I did. This little episode was only one of many that had caused delay and frustration in planning the Queen's visit to Expo. On my many visits there, when I had to discuss something with the manager of the Quebec pavilion, it had to be in French. Earlier in my career I would not have cared, but now my French was too rusty to do business in that tongue and I had to use an interpreter despite the fact that the manager spoke and understood English! Also, Daniel Johnson, the Premier, indicated to the Prime Minister that a visit by Her Majesty to Quebec apart from Expo would be fraught with difficulties. In any event, the Queen and the Prince stopped first at the United Kingdom pavilion and I never heard of any political repercussions.

279

My second day in London was spent mostly with Chevrier and he had me to lunch at his official residence in Brook Street with Sir Michael Adeane and others. We had useful discussions about the various programs. The next day I was invited to lunch with the Queen Mother at Clarence House, and I discussed with her the plan for visits to the Atlantic Provinces. It had been originally intended that in New Brunswick the Queen Mother would visit only Fredericton and Saint John in accordance with our policy. However, she expressed a desire to visit St. Andrews and the island of Campobello, where Franklin Delano Roosevelt had had a summer home that had been restored and maintained as a tourist attraction. These visits were arranged as she requested. Then, as one might expect, when Louis Robichaud, the Premier of New Brunswick, heard of the plans, he suggested that if she visited St. Andrews she should also visit the northern, i.e., the Acadian, part of his province and Moncton. As a result, a complicated change in arrangements had to be worked out whereby the Queen Mother would visit Moncton *after* she had been to Nova Scotia and Prince Edward Island.

At the luncheon in Clarence House, to which had also been invited Rear-Admiral Patrick Morgan, the master of the royal yacht *Britannia*, Sir Martin Gilliat, and other members of the household, the above plans were approved. Her Majesty had recently suffered a rather serious illness and I expressed the opinion that because of a twenty-four-foot tide at St. Andrews, she might find it distressing to mount or descend the many steps that would be necessary in landing and leaving both St. Andrews and Campobello. She said that by July she was sure there would be no problem, and the dear lady had the last word because, as she shook hands on my leaving, she said, "And don't worry about the tide, General. I'll practise running up and down steps in the meantime!"

In the late afternoon of one of my four days in London, after a long session with Major Clarke, Princess Alexandra's secretary, in his offices at Kensington Palace, I was driven to the Ogilvys' home in Richmond Park for an hour or two with the Princess and her husband to go over their plans. For some reason Major Clarke did not accompany me. It had been planned originally that their official visit would terminate at Winnipeg, on June 7, but she expressed a desire to visit Expo and to return to England by sea on the *Empress of Canada*, sailing from Montreal on June 9. Hence it was possible to arrange a stopover in Ottawa for a private lunch with the Governor General and Mrs. Michener and for a full-day visit to Expo. Happily, we were able to arrange a small private lunch with Madame Johnson, the wife of the Premier, at the Quebec pavilion.

The method of financing Alexandra's visit is of interest. As early as June, 1966, the Princess's comptroller wrote to Esmond Butler, Governor General Vanier's private secretary, reminding him that the Princess "is not a recipient of the Civil list" in the United Kingdom, seeking assurances that the Canadian government "would accept the full responsibility for all costs that are in any way attributable to the visit." In his reply Butler confirmed that the Canadian government would accept its responsibility.

During my visit to London, Major Clarke, acting as both secretary and comptroller, indicated to me that these costs would include, in addition to transportation and accommodation, the cost of additional clothing for the Princess (otherwise not required), the cost of small gifts she would feel obliged to give, including photographs, and a number of other miscellaneous tour expenses such as extra stationery and office supplies and insurance. For some of these items there could be no accounting. Since this was an unusual arrangement, but one I considered justified, I reported the facts to the Prime Minister immediately on my return to Ottawa and he gave approval in principle to this financial commitment. It was then a matter of determining the amount of money to be provided. On the basis of an initial estimate of 3,500 pounds, Major Clarke was given an advance of 2,000 pounds on the understanding that this would be accountable and that payment of any additional amount would only be made after presentation of an accounting of total expenditures. In the end, the total bill was 3,112 pounds. A curious side effect of this procedure was the insistence of the British treasury officials that Major Clarke should account to them also for this specifically Canadian grant or reimbursement of funds.

Another matter that required discussion was the number of persons who would accompany Princess Alexandra and Mr. Ogilvy. The government wanted the number to be kept to a minimum – the two principals, a secretary-comptroller, a lady-in-waiting, two maids, and a hairdresser. In addition to these, of course, would be my deputy, Ken McIlwraith, a Canadian transportation and accommodation officer, Lieutenant-Colonel Leonard, and an RCMP security officer, Inspector Pritchett. But the British, probably on the advice of Scotland Yard, proposed the addition of a British police officer and an air officer of the Queen's flight "in view of the considerable amount of air travel." Despite the delicate nature of these proposals, I firmly resisted them, contending that our own security arrangements were adequate and our air civilian and Air Force personnel were quite competent. Princess Alexandra and Mr. Ogilvy readily agreed with me and so it proved.

The official part of Princess Alexandra and Mr. Ogilvy's visit began in Toronto on May 14 and ended in Winnipeg on June 7. With the unofficial visits to Ottawa and Montreal, from June 7 to June 9, the total tour lasted twenty-six days. In that time the royal party travelled approximately eight thousand miles by air, rail, and road; visited eleven cities and towns (not including Ottawa and Montreal) in five provinces and two federal territories; and participated in over ninety programmed events, at thirty-five of which Princess Alexandra made short speeches. As Colonel-in-Chief of the Queen's Own Rifles, she spent an afternoon and evening with Honorary Colonel J.G.K. Strathy at his country home near Toronto, and she visited all three battalions of her regiment in Toronto, Victoria, and Calgary. Weather conditions were almost ideal throughout the tour, except for a day and a half in Calgary and part of the rest periods in Banff and Jasper. No changes had to be made in the itinerary, and no scheduled events had to be cancelled or postponed. Local press coverage of the visits in each area was substantially greater than anticipated and uniformly favourable. Television and radio coverage, thanks to the excellent preparatory work by my liaison officer, was both extensive and very effective. The Princess and Mr. Ogilvy made an excellent impression both on the hundreds of people they met and on the thousands more who saw them. It was agreed that this was a particularly successful royal visit.

My deputy, Ken McIlwraith, accompanied Princess Alexandra so that I might remain in Ottawa to continue working on the detailed program for the Queen and Queen Mother. I recognized from the beginning that security, particularly for the Queen, was perhaps going to be something of a problem. It must be recalled that at the time of her visit to Canada in 1964 (with which visit I had not been involved), there had been threats, written and spoken, and there had been anti-monarchist demonstrations in some places. There had been the assassination of U.S. President Kennedy in Dallas, Texas, and in 1966, in Belfast, the Queen's car was struck by a heavy object thrown from the top of a building. Knowledge of activities and threats to the Crown by small groups of separatists in Quebec led Commissioner George McLellan, head of the RCMP, to the opinion that there was a definite possibility of acts to embarrass the

Queen – protesters lying in the road, throwing paint, etc. – and even to cause her harm and place her life in jeopardy. Inspector Gordon Pritchett, my adviser on security, accompanied me and my deputy in all our planning trips and advised us on routes to be travelled, locations of platforms, seats in theatres, reception lines, and the like. While Pritchett was away with Princess Alexandra, the Commissioner himself worked with me on arrangements for the Queen for Ottawa and particularly for Expo. I knew from my discussions with all the royal visitors that they doubted the need for these stringent security plans. However, it was difficult if not imprudent to overrule the opinion and advice of my security experts. The issue came to a head over the use of an open car in Ottawa. George McLellan was insisting on the use of a special armoured limousine used during the Queen's visit in 1964, which for several reasons the Queen and Prince Philip abhorred. Finally, Commissioner McLellan, on a trip to London, consulted with officials at Scotland Yard and with the Queen's personal security officer, Chief Superintendent Perkins. As a result of these discussions, he agreed that an open car might be used in Ottawa only.

It may be of interest to know something of the escort and security duties performed by the Marine Division of the RCMP. These boats accompanied and were in constant radio communication with the royal yacht during all times that the Queen was aboard and during much of the time that the Queen Mother was aboard during her visit to the Atlantic Provinces. They also arranged for and co-ordinated the operations of frogmen who did underwater search at each of the St. Lawrence Seaway locks before transit by *Britannia*.

According to plan the Queen and the Duke of Edinburgh arrived at Uplands Airport, Ottawa, at 5:30 p.m., June 29. After the customary greetings, they had a quiet dinner at Government House and retired, it then being almost midnight London time. The next day was filled with the usual types of appointments, culminating in a state dinner.

Saturday, July 1, was the important day, Canada's Centenary. So that great numbers of people of all ages might witness this formal and historic occasion, it was decided to hold the ceremonies out of doors. Accordingly, a stage was erected on Parliament Hill with carpets, drapes, and the necessary furniture to resemble the interior of the Senate Chamber. The Queen and Prince Philip arrived from Government

House at 10:15 a.m., met the clergy, and the service began at 10:30 a.m. This was over by eleven o'clock, and the royal couple, with their attendants, mounted the stage and took their places. Her Majesty read her reply to the loyal addresses that had been read by the speakers of the Senate and of the House of Commons. The entire ceremony was completed by 11:30 a.m.

After a quiet luncheon there was a busy afternoon: at a children's picnic party on the lawn in front of the Parliament Buildings; another gathering, of teenagers, at Lansdowne Park; a short call at the Ottawa City Hall; then a garden party at Rideau Hall from 4:30 to 5:30 p.m. In the evening, the Queen and Prince dined with Prime Minister and Mrs. Pearson at their official residence and, following that, at 10:15 p.m., the whole party drove to Nepean Point, a high cape at the mouth of the Rideau River and across from the Parliament Buildings. From this vantage point on suitable stands, they witnessed a spectacular display of fireworks until almost midnight. And so ended the hundredth year of our country's birth.

Sunday, July 2, was rather a quiet day, attending morning service at Christ Church Cathedral, a quiet lunch, and then a two-hour drive through the countryside to board *Britannia* at Cornwall. During the night we had a pleasant voyage down the river to berth at a jetty close to the site of Expo.

July 3 was spent seeing the sights of this great exhibition and in the evening Her Majesty entertained some fifty guests at dinner on *Britannia*. The next day was a pleasant trip through the Thousand Islands to anchor off Kingston at 5:00 p.m. Here again, the Queen entertained almost fifty guests, including the premiers and lieutenant-governors of the provinces. July 5, the last day of the royal party's visit, was busy with engagements until the final dinner at Rideau Hall that evening for younger Canadians who had distinguished themselves in the professions, the arts, industry, athletics, and so on. Following the dinner, Queen Elizabeth and Prince Philip went directly to the RCAF station and, after bidding farewell to the notables, left for London.

On this visit the Queen again had as her Canadian equerry Colonel Roland (Roy) Reid of the Royal 22nd Regiment, who had been one of the three Canadian equerries in 1959. Roy Reid was exceptionally well suited for that appointment, always alert, intelligent, calm, pleasant, humorous, and much appreciated and admired by the royal couple.

Colonel Reid and I, as had been the case in 1959, stood at the foot of the aircraft steps to receive a warm handshake and thanks for our part in arranging this six-day visit.

The Queen Mother's visit to the Atlantic Provinces began on Monday, July 10, and ended on July 22, 1967. Arriving at Halifax by Air Canada from London, she and her party of twenty, including personal attendants and staff, transferred to an Andover of the Queen's flight and flew to Saint John, New Brunswick, where the royal yacht *Britannia* awaited them. I had been invited to travel with Her Majesty. The next day was spent in Saint John and on the 12th we flew to Fredericton, where she spent the morning. After lunch at the University of New Brunswick, there was a half-hour drive to Canadian Forces Base at Gagetown. Here she was greeted by a magnificent sight: three battalions of the Black Watch (Royal Highlanders of Canada) were drawn up on the parade ground with detachments from five cadet corps. The Queen Mother is Colonel-in-Chief of this fine regiment, and after her inspection she circulated among all ranks of the unit at a garden party. It was a beautiful day and I think the highlight of her visit to Canada. After the garden party, we flew back to Saint John and boarded *Britannia* for a night voyage down the Bay of Fundy to anchor off St. Andrews the next morning. This day, July 13, was spent at St. Andrews and Campobello Island, where Her Majesty was much interested in seeing the boyhood summer home of Franklin Delano Roosevelt. The Queen Mother and the late King George VI had been the American President's house guests at the White House in 1939.

During the night of July 13-14, *Britannia* carried us around the southern tip of Nova Scotia to Halifax, where she spent the day. During the following week, using the yacht most of the time, and motor cars occasionally, the party visited Antigonish, Sydney, Charlottetown, Summerside, and Moncton. The final two days, from 10:00 a.m. Thursday, July 20, to 10:00 a.m., Saturday, July 22, were spent at Saint John.

As had been the practice with the Queen, the dear Queen Mother bade me farewell at the foot of the aircraft steps with a handshake and warm words of thanks. Before we left the yacht, she had given me a signed and framed photograph and presented me with two beautiful sterling silver dishes with the royal cypher engraved on them. "These are for your wife," she said. A few weeks later, it was a pleasant surprise to receive a telegram, as follows: "Members of Queen Elizabeth's Household and officers of the Royal Yacht, reunited at Clarence House, send their best wishes to their mentor, guide, and friend." I greatly appreciated their thoughtfulness.

It was also a pleasure to receive from Prime Minister Pearson the following letter on September 18, 1967:

> I am writing to express my thanks and those of the Government to you for the splendid work you have done in the detailed planning and conduct of the program of Royal Visits in this Centennial Year. All of these visits have been outstandingly successful, and together they have formed a colourful and inspiring part of the overall program. In particular, the visit of Her Majesty and H.R.H. Prince Philip, and the events surrounding July 1st, provoked a memorable demonstration of national fervor and contributed directly toward the strengthening of national unity. Your own role in these events was of key importance and is greatly appreciated.
>
> I realize that it was a personal sacrifice for you to leave your life of retirement for this purpose, but I do hope that your satisfaction with the outcome will confirm to you, as it does to me, that your time was well spent in an important national service.

Before I left Ottawa and closed up shop, my staff gave me a farewell party and presented me with a pewter beer stein with the words "A Smashing Success" engraved thereon. Great credit must go to the many people who assisted in designing the program and in having it successfully carried through. This applies not only to office staff, security personnel, and transport and baggage people, but also to the workmen, decorators, caterers, and others who skilfully and cheerfully did their part. True, I was the one who had the responsibility of co-ordinating and approving or recommending approval by the Prime Minister and the royal visitors, and this did require in no small measure tact, understanding, sympathetic consideration of requests, and firmness where necessary. The year spent on this task was not easy but did give me some satisfaction in the belief that I had made a contribution to the commemoration of Canada's Centennial.

16
LOOKING BACK

And now as the shadows lengthen and I draw near to the end of life's long road, it is with deep sorrow that I must refer to a profound and distressing change in my life. A few years ago my darling Jean began to show signs of the onset of the dreaded Alzheimer's disease. This progressed until she had to be admitted to a nursing home about a year ago, where I now go to see her several times a week. Our good son Peter, still a bachelor, lives and works in Hamilton but comes to see me each weekend. Always cheerful, his visits are like a ray of sunshine, and together we go to see my wife, his mother. It is not exactly a happy experience, but at least for an hour or two we have a sort of family reunion.

So it is that I live along with the help and care of a housekeeper-attendant, Mrs. Joyce Nason, in my comfortable Oakville home overlooking Lake Ontario. Though my thoughts are often with Jean, who made the home so pleasant, I am fortunate, I suppose, in having many happy memories not only of her, but also of friendships made and tasks performed on a purely voluntary basis – often acting as a troubleshooter. There were the five years as president of the National Council of the Boy Scouts of Canada; the three years as member of the National Council of the Duke of Edinburgh's Award Scheme in Canada; several years as Honorary Colonel of my old regiment and founder of the Officers' Association; vice-president of the Empire Club of Toronto; on the executive of the Corps of Commissionaires in Toronto; chairman of the Oakville-Trafalgar Public Utilities Commission; an original member of the Halton District Health Council; an honorary governor of the Canadian Association for the Mentally Retarded; and other positions and activities. Truly, until about the age of seventy-five, there was little time for leisure.

A short time ago a friend asked me what I thought were the qualities that led me to the positions I have held in both the civil and military fields. I replied by saying that it was a hard question to answer, one that perhaps others could answer better than I. However, I have given the matter some thought, so perhaps a few words on my philosophy of life will not be amiss. Seven qualities are essential.

1. *Work and diligence*. Little can be accomplished without them.

2. *Loyalty, integrity, and honesty*. Others must have faith in you.

3. *Intelligence and wisdom*. Continue to learn, but be not afraid to ask questions, apply knowledge wisely, and use common sense.

4. *Understanding and sympathy*. Try to see the other person's point of view, but do not necessarily agree with it.

5. *Determination*. Don't give up too easily. (I remember my dear old grandmother saying to me, when I had quit trying to tie a shoelace, "Now, now, Howie, don't give up so easily just because it's hard to do!")

6. *Gratitude*. This is the sign of good manners. How often have I done a favour or written a note of congratulations to someone and never received an acknowledgement?

7. *A sense of humour*. Applied at the proper time, in the proper place, this gift is a helpful attribute.

These seven qualities are the virtues I have tried to practise, perhaps not always with success.

I believe I have been lucky in life. It has been my good fortune never to be unemployed. There always seemed to be at least one job awaiting me. It has been said that successful commanders in time of war must not only be capable and efficient but must also be lucky. Certainly, good fortune smiled upon me.

And now, as I look back upon these events in a very full life, I lay aside my pen, close the book, and go forth to trim my roses. They badly need it!

EPILOGUE

I knew Howard Graham for many years. When I was a junior officer in England, he commanded the Hastings & Prince Edward Regiment. Later I served as his brigade major when he was in active command of the 1st Canadian Infantry Brigade in the Sicilian and Italian campaigns. I followed in his footsteps to command the Hastings & Prince Edward Regiment in Holland and eventually, like him, to become its Honorary Colonel.

In spite of differences in age and rank, we became friends and kept in contact through the years. Recently he asked for my help in publishing his memoirs. One does not argue with generals. Eventually, I became his literary executor and my friend, Jack McClelland, his publisher. This project would not have been possible, however, had not John Robert Colombo become his editor; tactful, persistent, and able, he gained the General's confidence and got the job done.

The present book is the result. It is vintage Graham. The original manuscript was written in the General's own hand without editorial interference of any kind. The final copy, the result of an intense campaign between the General and Colombo, was approved by the General, and though abbreviated in length, its content and style remain unchanged. He was in command to the end!

Howard Graham was a man of humble origins who became one of Canada's most distinguished soldiers, a man who was equally at home in the field with his troops, in the law courts of Ontario, in the boardrooms of business, or on the royal yacht *Britannia* with the royal family.

He was truly a great Canadian!

He died on September 28, 1986, at the age of eighty-eight.

Colonel George E. Renison, D.S.O., E.D., U.L.,
Honorary Colonel of the Hastings & Prince Edward Regiment
Toronto
January 1, 1987

INDEX

Buckingham Palace, 275-76
Bucknell, Colonel, 105
Buffalo, 9-10
Buffalo Street Railway, 9, 10, 11
Bunche, Mike, 233
Bunche, Ralph, 233
Burke, Arthur, 96
Burma, 128
Burness, Major, 105
Burns, Lt.-Gen. E.L.M. "Tommy,"
    107, 232, 233, 234-35
Burtt, Ross, 254
Butler, Esmond, 247, 251, 252, 284

Cadet Corps, Ont., 41, 43, 44
Calabrian peninsula, Italy, 183-84
Calais, 115
Calgary, 253, 282
Caltagirone, Sicily, 157, 164
Camblain L'Abbe, France, 62
Cambrai, France, 274
Cameron, Rev. W.A., 76
Campbell, Air Marshal, 237, 238
Campbell, Maj. Alex, 152
Camp Borden, 222, 223, 224, 226
Camp Ipperwash, Ont., 223
Campney, Ralph, 224, 226, 227, 232,
    236, 237, 243
Campobasso, Italy, 187-88
Campobello Island, 280, 285
Canada Forces Act, 221
Canada House, London, 93, 209
Canada Steamship Lines, 256
Canadian Bar Assoc., 91
Canadian Club, 84, 271
Canadian Corps, 57, 59-66, 69-70
Canadian Depository for Securities
    Ltd., 265-66
Canadian General Hospital: Bari
    (No. 1), 192-93; Basingstoke (No.
    4), 70-71; Caserta (No. 14), 193;
    London (No. 10), 193
Canadian Military Cemetery, Nijme-
    gen, 214
Canadian National Exhibition, 63
Canadian National Railway, 221, 249

Canadian Pacific Railway, 38, 39, 73,
    221
Canadian Press, 254
Canadian Scottish regiment, 135
Canadian Special Service Force
    (CASF), 218, 220-21, 230-31
Cannes, 68-69
Canterbury Belle, 118-19
Carlton and York regiment, 183
Carson, Cyril, 79
Cartwright, John, 79
Cassino, Italy, 275
Castropignano, Italy, 189
Castrovillari, Italy, 186
Catania, Sicily, 164, 177, 184
Catanzaro, Italy, 186
Catherine Street School, Belleville,
    51, 52
Cayuga (destroyer), 219
Cenotaph, Whitehall, 103
Central Command (Canadian Army),
    222-27
Centre Island, Toronto, 77
Century Bank of New York City, 86
Charing Cross Road, London, 60
Charleroi, Belgium, 70
Charlottetown, 285
Château Laurier, Ottawa, 94
Chemical Control Board, 135
Cherbourg, France, 92
Cherkley Court, Mickelham, 212
Chevrier, Lionel, 277, 278, 280
Chicago, 247, 253
Chichester, England, 125
Chorley Park Military Hospital,
    Toronto, 203
Christie, Harry, 72
Churchill, Winston, 66, 115, 119,
    120-21, 123-24, 128, 136, 198, 246
Chrysler, 249
Civil Defence Organization, 243
Clarence House, London, 280
Clark, Lt.-Gen. S.F., 243
Clarke, Major, 280, 281
Clarkson, Ont., 91
Claxton, Brooke, 208, 219

Quebec City, 91, 199, 253
Queen Elizabeth (liner), 144
Queen Mary Hospital, Montreal, 226
Queen's Hotel, Toronto, 73
Queen's Own Rifles, 133, 233-34, 282

Ralston, Colonel, 197-98, 199-200, 201, 202
Rank, Mrs., 130
Rank, J. Arthur, 130
Rankovic, Bozo, 86
Rawlinson, Gen. Sir Henry, 63
RCMP, 249, 252, 281, 282-83; Marine Division, 283
Red Star Line, 56
Regalbuto, Sicily, 165, 177
Reggio di Calabria, Italy, 183, 185
Regina, 253
Regina Rifles, 135
Reid, Col. Roland, 284-85
Renison, Col. George E., 289-90
Rennie's catalogue, 24
Renown (battle cruiser), 111
Rhyl, Wales, 71
Richards, Group Capt. Gordon, 250
Richmond, Duchess of, 132
Riddell, W.R., 78
Roberts, A. Kelso, 79, 259, 271
Roberts, Maj.-Gen. J.H., 140
Robertson, Fraser, 273
Robertson, Norman, 93, 211, 213, 214
Robert the Bruce, 60
Robichaud, Louis, 280
Robinette, John J., 258, 271
Rockcliffe Park, Ottawa, 228
Rockingham, Brig. John, 220
Rocky Mountains, 253
Roosevelt, Franklin Delano, 215, 280, 285
Ross, Rev. Geordie, 44
Rotary Club, Trenton, 84
Rowell, Newton Wesley, 84-85
Royal Air Force (RAF), 131, 139, 211
Royal Artillery, 166
Royal Canadian Air Force (RCAF), 88, 199, 210, 215, 216, 219, 230, 234, 237, 249, 250, 251, 254, 260
Royal Canadian Dragoons, 236
Royal Canadian Engineers, 230
Royal Canadian Legion, 89, 102, 274
Royal Canadian Navy, 131, 141, 142, 143, 148-49, 151, 185, 210, 215, 219, 237, 249, 250
Royal Canadian Regiment (RCR), 112, 113, 115, 119-20, 130, 132, 136, 141, 150, 151, 152, 157-58, 159, 165, 166, 172, 175-76, 177, 182, 188, 191-92
Royal Marines, 142
Royal Military College, Kingston, 104, 140
Royal Newfoundland Regiment, 274
Royal Sussex Regiment, 125-26
Royal Troon golf course, 143
Royal 22nd Regiment, 183, 284
Royal Winnipeg Rifles, 135
Royal York Hotel, Toronto, 73, 272
Rundstedt, Commander, 133
Ryde, Township of, 90

St. Andrews, N.B., 280, 285
St. Andrews Church, Toronto, 76
St. Hubert Airport, Que., 252
Saint John, N.B., 280, 285
St. Julien, 44, 112
St. Lambert, 253
St. Laurent, Louis, 91, 219-20, 237, 242
St. Lawrence Seaway, 242, 246, 247-48, 253, 256-58, 283
St. Paul's Cathedral, London, 213
St. Peter-in-Chains Roman Catholic Church, Trenton, 35
St. Pol, 61, 67
Salerno, Italy, 186-87
Salisbury Plain, 114, 210
Salmon, Maj.-Gen. Harry, 112-14, 115, 120-22, 124, 126, 134, 135, 136, 137, 138, 139-40
Salvation Army, 119
San Leonardo, Italy, 190, 191
Santa Teresa, Sicily, 184